Essentials of
the BRAIN

AN INTRODUCTORY GUIDE

RUDOLPH C. HATFIELD, PhD

New York

This book is dedicated to my family
for putting up with my antics for all these years.

FALL RIVER PRESS

New York

An Imprint of Sterling Publishing Co., Inc.
1166 Avenue of the Americas
New York, NY 10036

Fall River Press and the distinctive Fall River Press logo are
registered trademarks of Barnes & Noble, Inc.

© 2013 by F+W Media, Inc.

All rights reserved. No part of this publication may be reproduced, stored in a retrieval
system, or transmitted in any form or by any means (including electronic, mechanical,
photocopying, recording, or otherwise) without prior written permission from the publisher.

ISBN 978-1-4351-6518-2

For information about custom editions, special sales, and premium and corporate purchases,
please contact Sterling Special Sales at 800-805-5489 or specialsales@sterlingpublishing.com.

Manufactured in Singapore

4 6 8 10 9 7 5 3

sterlingpublishing.com

Images by Eric Andrews

Contents

ACKNOWLEDGMENTS

I have many people to thank for contributing to the information in this book. None of them have contributed directly, but instead have influenced or taught me over the years. I would like to acknowledge all my professors at the University of Michigan and Wayne State University for their wonderful teaching and patience with me as a student so long ago. However, I really need to thank the thousands of patients with brain injuries, neurological disorders, psychiatric disorders, and other brain-related problems whom I have assessed or treated over the years. It is from you that I have learned more than I ever could have learned in the classroom. God bless you all.

THE TOP 10 FASCINATING FACTS ABOUT YOUR BRAIN

1. Your brain is forever changing. The ability of the brain to respond to experience and physical change is called *plasticity*.

2. You have holes in your brain. There are four spaces in the brain called *ventricles* that protect the brain from extreme pressure, such as that caused by head trauma.

3. Your brain is the original 3-D. Your eyes transmit small, inverted two-dimensional images to your brain, and your brain fills in missing information and creates three-dimensional images.

4. Your working memory can only hold about five to nine "chunks" of information, and its typical duration is about thirty seconds.

5. The perception of pain is paradoxical. In certain circumstances your brain can actually block the perception of pain and allow you to continue with an action.

6. Everyone dreams, even those people who claim they don't. Dreams are an attempt by your brain to make sense of the mass of information sent to it during REM sleep.

7. The frontal cortex displays the most prolonged development of all brain areas, not becoming mature until adolescence.

8. Attention is among the most sensitive of cognitive abilities. Difficulties with attention and concentration are some of the most common problems associated with disorders such as depression or anxiety.

9. Only a small percentage of all brain activity actually reaches conscious awareness.

10. Your memory is not a tape recorder. When you recall memories, your brain must actually recreate them. This recreation is subject to outside influences (context, suggestion, emotional states, etc.) and can alter the memory of past events.

INTRODUCTION

THE 1980s WERE DESIGNATED as "the decade of the brain." At that time fields like cognitive science, neuroscience, and other areas of brain-related study were flourishing, and it was believed that the period between 1980 and 1990 would reveal the inner workings of the human mind. Well, that decade has come and gone and advances in understanding the brain have not ceased. Looking back on those years, it seems like the more that scientists learned, the more they realized that they did not understand. Science has not even begun to comprehend the brain. That makes writing this book all the easier because, in reality, each chapter could simply state, "No one really understands how the brain does this," and this would be the most accurate book on brain-functioning ever written.

Nonetheless, science has made some exciting progress in understanding how the brain may work. Unfortunately, much of this progress can be lost on those who do not have extensive backgrounds in biology, anatomy, neuroscience, and psychology. While it is necessary to have a very basic knowledge of these fields in order to understand the brain, it certainly is not necessary to be an expert in any one of them. This book attempts to allow someone without expertise in neuroscience to understand how her brain works and how the brain organizes sensation, perception, thinking, and feeling.

The human brain is the most complicated entity (for lack of a better word) in the entire universe. The human brain has the capacity to perform functions that, as far as science knows, cannot be performed by any other animal or by any machine. This book attempts to explain the basic functions of the brain without using complicated computer analogies and endless neuroanatomical drawings and pictures. Computer models can be useful in understanding how neural networks signal and code information, but this book does not attempt to provide a reductionistic viewpoint of behavior. Instead, this book looks at the top-down processes of brain functioning. References allowing for more in-depth examinations of topics are provided at the end of the book, and there is a glossary at the end of the book with links to websites that enhance the book material. The motivated reader will hopefully investigate these.

A QUICK VIEW

IT IS VERY DIFFICULT to begin a book on the brain without sounding like a cliché. The brain is certainly the most complex known entity in the universe. It is more complex than all of quantum physics (which was created by someone's brain), all of the laws of the universe, and any other phenomenon that you can think of. This first chapter will look at some fundamental concepts that need to be addressed before discussing the brain in more detail. It will start with very basic information so that you can develop an overall understanding of how the brain functions.

The Two Different Nervous Systems

There are two major divisions of the nervous system: the central nervous system and the peripheral nervous system. The *central nervous system* (CNS) consists of the brain and the spinal cord. The *peripheral nervous system* (PNS) consists of nerves outside of the brain and the spinal cord. The peripheral nervous system is traditionally divided further into two groups: the somatic nervous system (generally considered to be under voluntary control, such as the skeletal muscles) and the autonomic nervous system (generally considered not to be under voluntary control, such as digestion). The autonomic nervous system is further divided into the sympathetic nervous system (which functions to speed up your body's organs) and the parasympathetic nervous system (which functions to slow down your body's organs). This book will concentrate on the brain and its interaction with other nervous systems and the body.

> The **enteric nervous system** is a third division of the autonomic nervous system not often mentioned in many texts on the brain. The enteric nervous system is a network of nerves that innervate the **viscera**, organs in the body cavities, especially in the abdominal cavity (e.g., the gastrointestinal tract, pancreas, gall bladder, etc.).

Brain Basics

The average human brain weighs about three pounds and is the consistency of Jell-O (the pickled brains you see in jars are actually hardened). However, there is quite a bit of variation in brain size just like there is variation in body size. A person with a bigger brain is not necessarily smarter than one with a smaller brain, all other things being equal. For instance, Albert Einstein's brain, which he donated to science after his death, is reported to have weighed only 1,230 grams, or about 2.71 pounds, which is slightly smaller than average.

> A large number of brain cells are lost through attrition, programmed cell death, and other causes. However, claims that 5,000–10,000 or more brain cells are lost daily are unfounded. No one really knows how many brain cells there are and certainly no one knows how many get "lost."

Many texts report that the human brain contains about 100 billion nerve cells (neurons) and trillions of support cells (e.g., glial cells). However, more recent estimates have suggested that this figure is somewhat overstated. Neurons are nerve cells that are specific to the CNS and are connected in a number of intricate pathways and networks. The actual number of these connections may exceed 100 trillion! It is the connections between the neurons (the nerve cells in the brain) that allow neurons to communicate with each other, and this activity is responsible for all of your actions.

How the Central Nervous System Works

For most of the voluntary actions that people make (and a good number of involuntary ones), these initial behaviors begin in the brain where they are formulated. The message is then sent down the spinal cord into the peripheral nervous system allowing one to take action. Your central nervous system operates as a type of body control center and complex communication system that is composed of a sophisticated network operating both chemically and electrically. Your brain also responds to information that is transmitted from your sense organs through your spinal cord and relayed to your brain.

Incoming information is transmitted via *afferent* (incoming) nerve cells in sense organs to afferent neurons on the underside of your spinal cord (the *ventral*, or belly, side). This information is sent through the spinal cord to your brain. Your brain then interprets this information and the appropriate action is decided on. This response is sent via outgoing (*efferent*) nerve cells or neurons back down your spinal cord to your muscles (or whatever part of the body that is appropriate) via the *dorsal* (back) side of your spinal cord.

So for instance, if you are touching a soft fur, the information about the feel of the fur is sent from your skin to your spinal cord (via afferents) to your brain. Suppose you decide that it is pleasing and that you want to stroke it further (this decision takes place in your brain). That information is sent from your brain (via efferent nerve cells) to your spinal cord and then to the muscles in your arm and hand that allow you to stroke it. Your nervous system integrates, detects, and processes countless bits of information at any given moment.

There are situations when the brain is not involved in movement. Certain reflexes like the patellar reflex, when the doctor strikes your knee with a rubber mallet and your knee extends, do not involve your brain. These occur via a loop from the receptors in your body to your spinal cord and back again. However, for the vast majority of actions, the brain is in control.

Mixing Chemicals and Electricity: The Neuron

The main architect of everything that happens in your brain is a very special nerve cell called the *neuron*. Neurons come in many different shapes and sizes and there will be more concerning them in subsequent chapters of this book. The first order of business is to take a look at a typical neuron, and discuss its parts and how it basically works. **Figure 1-1** is a depiction of a typical neuron.

Neurons consist of several parts: At the top part of the neuron in **Figure 1-1** there are several structures known as *dendrites*. Dendrites receive chemical messages from other neurons. Moving down the neuron is the *soma* or cell body. Here all the functions needed to maintain the health and integrity of the neuron occur, such as metabolic functions and so forth. Moving further down the

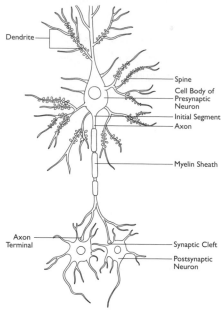

Figure 1-1: A Typical Neuron

neuron leads to the *axon*, which is the signaling part of the neuron. Most axons are covered with a fatty sheath known as the *myelin sheath*; however, the entire axon is not covered with myelin and there are small areas where the axon is uncovered. The myelin sheaths resemble elongated pillows running down the length of the axon (these spaces in between the myelinated areas are termed the *nodes of Ranvier*). At the end of the axon there is a bulb where the axon terminates (the *terminal bulb*) and a space called the *synapse* that separates the axon of one neuron (the sending part) from the dendrites of another neuron (the receiving part). The neuron depicted in **Figure 1-1** is a prototype; there are several different types of neurons. In Appendix B you will find a link that allows you to view actual neurons.

> ***Excitatory neurons*** stimulate neuronal firing, whereas ***inhibitory neurons*** reduce the rate of neuronal firing. ***Motor neurons*** are involved in motor functioning, whereas ***sensory neurons*** are involved in detecting and interpreting sensory stimulation. An ***interneuron*** connects other neurons together and is neither sensory nor motor in its functioning.

How Neurons Communicate

The process of signaling between neurons is quite complicated and will be simplified for this discussion. Basically what happens is that stimulation from sensory systems or from your thoughts results in a neuron being "activated." Typically this consists of chemical substances known as *neurotransmitters* attaching themselves to the dendrites of a neuron. If a sufficient amount of neurotransmitters attach themselves to the neuron, this will result in the activation of an electric charge (an actual signal) known as an *action potential* being sent down the axon of the neuron.

The process of the action potential in a neuron depends on its capacity to react to a stimulus with an electrical discharge. This process is quite complicated but it involves changes in the electrical charges of the ions within the neuron compared to the electrical charges of the ions outside of the neuron's cell wall. When the neuron is activated, the negative-based inner charge of the neuron becomes more positive due to an exchange of positively charged ions through the cell membrane via special gates. This results in a loss of negatively charged ions, which move out of the cell (when the neuron is not stimulated, this process is stabilized to maintain a relatively consistent balance). This change in the charge of the neuron results in the action potential, or electrical charge, being generated and traveling down the axon of the neuron.

The myelin sheath on the axon acts as a sort of insulator to facilitate the transmission of the electric charge. The electric charge literally jumps from space to space (the unmyelinated nodes of Ranvier) as it moves across the axon. When the electric charge reaches the end of the axon (the *presynaptic terminals*), it results in the release of neurotransmitters into the synaptic space. These neurotransmitters will attach themselves to the dendrites of adjoining neurons and the process will continue. Once the neuron has released its load of neurotransmitters, it is in a refractory period for a short time until the cell normalizes.

Most neurons communicate in this manner, but there are a very small number of neurons that communicate with one another via electrical charges. These neurons are called *gap junctions*.

When a neuron connects with another neuron, it is said to **synapse** at that site. Most of the neurons in the brain communicate chemically with each other, as this allows for a wide variation in the types of messages that can be sent and a greater potential to mediate the signal firing and signal strength than if neurons communicated via an electrical impulse.

Every neuron is only activated by a specific neurotransmitter or a specific set of neurotransmitters. Thus, the signaling *between* neurons is accomplished via a chemical process (neurotransmitter), whereas the messaging *within* the neuron is accomplished via an electric charge (action potential). These chemical communications between neurons and the communications they deliver result in your thoughts, feelings, and actions.

Neurotransmitters

Neurotransmitters are chemical substances that the brain uses for communication between neurons and are created within neurons. There are hundreds of chemical substances that qualify as neurotransmitters. These substances are not used exclusively in the brain, although when they are utilized in the brain, they are called neurotransmitters. Outside of the brain when the same substances travel through the bloodstream, they are referred to as *hormones*. For example, the neurotransmitter serotonin, known to be important in mood, is also important in the process of digestion and is found in the gut. Most neurotransmitters have multiple functions. The neurotransmitter dopamine has important functions regarding your mood, movements, and memory.

> The movement of the action potential down the axon of a neuron is one-directional. An action potential travels down the neuron toward the synapse—never the other way around. Once an action potential traverses the nodes of Ranvier, the upstream portion of the cell enters a refractory period and cannot generate a new action potential for a short period of time.

Neurons fire on an "all or none" basis; that is, there is no such thing as a neuron firing halfway (there is no such thing as half of an action potential). Neurotransmission is regulated by how often, how fast, or by the pattern of neuronal firing. Some neurotransmitters excite the system (cause a more rapid firing of neurons); some inhibit the system (cause neurons to fire at a slower pace); and others modulate how neurons fire. It is through these patterns of firing that the different signals are sent back and forth in the brain and to the body. The following table lists several of the more common neurotransmitters. You can also find a link to a site in Appendix B that can provide you with more information about neurotransmitters.

Several Well-Known Neurotransmitters

Neurotransmitter	Basic Function
Dopamine	Responsible for arousal levels, mood; important in motivation; involved in voluntary movements
Serotonin	Effects on mood and anxiety, appetite, sleep, memory and learning, temperature regulation, and other functions
Acetylcholine	Controls activity in brain areas associated with attention, learning and memory, movement, and other functions
Epinephrine (Adrenaline)	Effects on attentiveness and mental focus; outside the CNS, it is involved in the "fight-or-flight response"
Glutamate	The major excitatory neurotransmitter in the brain
Enkephalins and Endorphins	Modulate pain, reduce stress, and promote a sensation of calmness; related to opiate drugs like heroin, they also decrease physical functions such as respiration
GABA (Gamma-aminobutyric acid)	The major inhibitory neurotransmitter in the CNS, it helps neurons recover after transmission, reduces anxiety, and reduces stress

Other Types of Cells in the CNS

Neurons are not the only cells in the brain and spinal cord. There are other cells such as glial cells, which perform a number of functions in the brain. At one time it was believed that glial cells were just support cells; however, it is now known that glial cells perform other functions. These functions include surrounding and holding neurons in place, supplying nutrients and oxygen to neurons, insulating one neuron from another neuron, forming myelin, and destroying and removing dead neurons. Three types of CNS glial cells are astrocytes, oligodendrocytes, and microglia. While there are billions of neurons in the brain, there are one to five *trillion* glial cells in the brain, or nearly ten to fifty times more glial cells in the brain than neurons!

Astrocytes perform numerous functions. They support the endothelial cells (a thin layer of cells lining the interior of blood vessels) that form the blood-brain barrier, provide nutrients, and help repair the brain and spinal cord following trauma. The major function of the *oligodendrocytes* is to provide support for axons and to produce the myelin sheath that insulates them. *Microglia* are the immune defenses in the CNS constantly hunting for damaged neurons and infectious agents.

The Spinal Cord

An extremely important part of the CNS is the spinal cord. Without the spinal cord your brain would be totally useless. It is important to remember that the brain cannot transmit information or receive any information without the spinal cord. The human spinal cord is about 18 inches in length (about 44 to 45 cm, but it is typically slightly longer in men than in women). The spinal cord is housed by the spinal vertebrae that offer some protection from injury. Neurons project from the spinal cord to other areas of the body, and nerves from the body project to the spinal cord.

> Cranial nerves emerge strictly from the brain, as opposed to spinal nerves that project from segments of the spinal cord. In humans, there are twelve pairs of cranial nerves that are involved in different functions. Ten of the pairs of cranial nerves project from the brain stem, whereas two pairs project from the cerebrum.

When the spinal cord is transected (cut) or damaged, the brain cannot communicate with regions of the body below the site of the cut. There are three major functions of the spinal cord:

1. Acting as an agent for transmitting motor information, which travels from the brain, down the spinal cord, and to the body

2. Acting as an agent for the relay of sensory information, which travels from the body to the brain

3. Acting as a center for coordinating some types of reflexes

Neuroimaging and Other Ways of Looking at the Brain

Structural and functional brain imaging has revolutionized the fields of neuroscience and medicine. The American neurosurgeon Walter Dandy first introduced ventriculography and later developed pneumoencephalography, both imaging methods in the early 1900s that often required the pumping of air into the brain; however, both procedures carried significant risks and could be quite painful. The technique of cerebral angiography was introduced in the late 1920s by neurologist Egas Moniz. This is a technique that takes pictures of the veins and arteries in the brain. This technique became refined and is still an important tool that is used today in neurosurgery.

> Brain imaging can be noninvasive, such that no penetration of tissue occurs (CT or MRI), or invasive, where there are such tissue penetrations (PET, fMRI). The brain is never penetrated. **Invasive** images sound threatening, but there typically is no more than minimal discomfort involved as only an IV is used.

Further advancements, such as computerized tomography (CT), magnetic resonance imaging (MRI), and positron emission tomography (PET), have led to researchers and physicians being able to visualize the brain and other areas of the body for diagnostic, treatment, and research purposes. The advancements in neuroimaging have led to many remarkable discoveries. CT scans literally take x-rays of brain tissue, whereas MRI scans use a completely different technique that results in a much more detailed image of the brain tissue. These two techniques are known as *structural imaging techniques* in that they take a static picture of the brain and can be used to determine changes in brain structure. PET scans are a method whereby researchers can view the metabolic changes in the brain, thus PET scans and other scans like them such as functional MRI (fMRI) are known as *functional imaging techniques*. During a PET scan, a harmless radioactive isotope is injected into the bloodstream and tracked to allow researchers to study brain metabolism.

Electroencephalography (EEG) is the recording of electrical activity of the brain by placing electrodes along one's scalp. Physicians who specialize in performing and interpreting neuroimaging techniques are called *neuroradiologists*.

There are a number of other advanced neuroimaging techniques and new developments in the field of imaging the brain being made all the time. There are several links provided in Appendix B to allow you to see how these techniques appear.

DIVIDING UP
the Brain

IN ORDER TO UNDERSTAND cognition, emotions, and other facets of experience, you must first understand the structure of the brain and how its parts work. This chapter will introduce the basic divisions of the brain and their associated functions. One thing to keep in mind is that different neuroanatomy texts may differ in description of what components make up a particular structure. This book depicts brain structures as most anatomists describe them.

North-South-In-Out: Directional Descriptions

The first thing to understand is the directional format that is used when describing the brain. Anatomists use specific terms to describe the front, the back, the top, the bottom, the sides, and inner structures of the brain. For example, the dorsal (back)/ventral (belly side) distinction was discussed previously during the discussion of the spinal cord. These distinctions can also be made in the brain. A brain area/structure located toward the top of the brain or above another structure is typically depicted as being *superior* to that structure, whereas something that is more toward the bottom or below

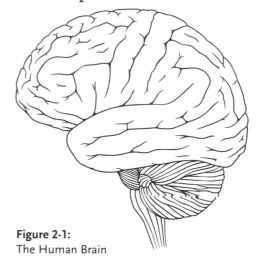

Figure 2-1:
The Human Brain

another structure is typically depicted as being *inferior* to it. In general when discussing the human brain, the terms *dorsal* and *superior* are equivalent, and *ventral* and *inferior* are equivalent terms. When neuroanatomists speak of a structure located on the side of the brain, they use the term *lateral*, whereas structures that are closer to the midline of the brain (toward the middle of inner portions) are considered to be *medial*. Moving toward the front portion of the brain (the front of your skull) is often referred to as being *anterior* (the older designation for this is *rostral*), whereas moving toward the back of your head is referred to as being *posterior* (older books may use the term *caudal*). The rostral/caudal designation is less confusing in animals that walk on four legs and have a tail but does not translate well in humans, and many often find it difficult to visualize. To simplify things, this book will attempt not to use such terms without explaining them. **Figure 2-1** depicts a typical brain.

Likewise, there are different perspectives that one can take when viewing the brain. **Figure 2-2** depicts the different perspectives of the horizontal plane, sagittal plane, and coronal plane designations that allow researchers and clinicians to view the brain from different angles.

Forebrain, Midbrain, Hindbrain

Some anatomists prefer to divide the brain into three sections: the forebrain, the midbrain, and the hindbrain. This distinction allows for the identification of more recently

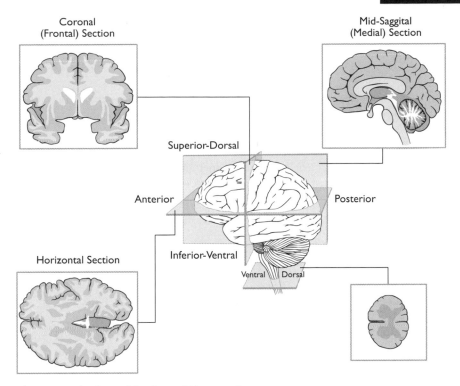

Coronal
(Frontal) Section

Mid-Saggital
(Medial) Section

Superior-Dorsal

Anterior

Posterior

Horizontal Section

Inferior-Ventral

Ventral   Dorsal

Figure 2-2: The Axis of the Central Nervous System

developed brain areas in terms of evolution and is consistent with higher order and lower order functions. The forebrain would be responsible for higher-order cognition, whereas the midbrain and hindbrain are responsible for more remedial or life-sustaining functions and allow for transitioning between the brain and spinal cord. These three categories divide the brain from the inner to outer sections.

- The **forebrain** is the most outward portion of the brain, consisting of the cerebrum, the limbic system, basal ganglia, the thalamus, and the hypothalamus (the forebrain may also be listed in texts as the telencephalon and diencephalon). The cerebrum refers to the lobes of the brain and the cortex of the brain. The limbic system is made up of the hippocampus and the amygdala, structures that are important in memory and emotion. The basal ganglia are composed of several structures that facilitate movement and aspects of cognition. The thalamus acts as a relay station that categorizes all sensory information entering the brain (except for the sense of smell) and then sends this information on to the appropriate areas of the brain. The hypothalamus is located just below the thalamus and is important in a number of functions, such as eating, drinking, sexual behavior, and hormone regulation. Collectively, the thalamus and hypothalamus are sometimes referred to as the *diencephalon*.

- The **midbrain** (mesencephalon) serves to relay information between the hindbrain and the forebrain, particularly for visual and auditory information.

- The **hindbrain** (myelencephalon) is the lowest portion of the brain and includes the pons, the cerebellum, and the medulla. The pons serves as a bridge toward the midbrain and has important functions in arousal and sleep. The cerebellum is a very important brain structure that is involved in coordinating body movements and in aspects of cognition. The medulla is the point where the brain connects to the spinal cord. This structure is important in regulating breathing and other automatic body functions.

What's with All the Wrinkles? Gyri and Sulci

The brain is covered with the cerebral cortex (which means bark or hide). The cortex has a convoluted look that is prevalent with "wrinkles" that cover the exterior of the brain. These wrinkles are referred to as *gyri* (*gyrus* for the singular). The cracks are referred to as *sulci* (*sulcus* in the singular). Some sulci run very deep in the brain, and these sulci are known as *fissures*. This arrangement of wrinkles and cracks allows a much broader surface area of tissue to be packed into a relatively small space in much the same way you can take a piece of paper, crumble it up into a ball, and fit it into a much smaller container than you could if it were left flat. This maximizes the functional area of your brain without giving you a head the size of a small end table.

There is quite a bit of variation in the appearance and length of the gyri and sulci and many of these are well described. Some gyri or sulci also separate lobes of the brain from one another or have different functions; however, in many cases they do not. The pattern of sulci and gyri in the human brain demonstrates quite a bit of variation between people and also between the two halves of any individual brain.

It is a myth that learning something new adds wrinkles to the brain. A fetus starts out with a smooth brain, but the gyri develop as the fetus develops. The wrinkles or gyri in the brain are distinct anatomical structures, relatively consistent across individuals, each with a designated name and location.

Lateralization: Left Brain or Right Brain

One of the most prominent distinctions in brain anatomy and brain functioning is the distinction of the cerebral hemispheres. These hemispheres make up the two halves of the brain: the left hemisphere (or "left brain") and the right hemisphere (or "right brain"). The two halves of the brain make up the entire brain and are distinctly separated by the longitudinal fissure, which is the deep groove in the center of the brain. Each hemisphere demonstrates some specialization of functions. For instance, the left hemisphere specializes in language, verbal memory, and logical thinking, whereas the right hemisphere appears to be more specialized for visual-spatial functions, nonverbal functions, understanding humor, and is involved in more intuitive types of processes. In addition, motor and sensory functions are lateralized such that the left side of the brain controls the right side of the body, and the right side of the brain controls the left side of the body. The advantage of brain lateralization appears to be that it allows the organism to be able to perform different tasks simultaneously. Other theories for brain lateralization include:

1. The *analytic-synthetic theory*, which surmises there are two basic modes of thought: analytic and synthetic modes. These have become segregated as a result of evolution and assigned to the left and right hemispheres, respectively. The problem with this theory is that it is a bit vague; it really is not possible to determine tasks that are purely "analytic" or "synthetic."

2. An outgrowth of the analytic-synthetic theory is the *linguistic theory*, which states that the primary role of the left hemisphere is language. Supporters of this theory point out that people who use sign language also appear to have left hemisphere –based language (for most people, language functions are located in the left hemisphere). However, there is no reason offered as to why the left hemisphere should be language dominant.

3. The *motor theory* of cerebral lateralization states that the left hemisphere is involved in the control of fine movements (of which language is one), and the right hemisphere is specialized for broader or gross movements. However, this theory does not suggest why these motor functions became lateralized in the first place.

Connecting the Two Halves and the "Split Brain"

Even though the brain is lateralized, both hemispheres are connected by tracts of nerve fibers known as *commissures*. These commissures allow the two halves of the brain to communicate with one another. The most prominent commissure is the *corpus callosum,* which is a large tract of neuron fibers that connects the left and right hemispheres of the brain. The corpus callosum has been the focus of important research allowing scientists to understand brain functions.

Patients with epilepsy have provided science with some interesting findings regarding the differences in the two hemispheres of the brain, based on a surgical procedure to treat their disorder that involves severing the corpus callosum. Epilepsy is a brain condition that results in seizures. Typically, an area of the brain is damaged or dysfunctional (either from birth or via some traumatic event) and the neurons in that area occasionally fire erratically. This leads to the erratic message being sent throughout the brain and across the corpus callosum, affecting the entire brain. When this happens people can have severe seizures. In some cases these seizures may not respond well to medication. In the 1960s, a surgical procedure called a corpus callosotomy, where the corpus callosum was cut, was used to control these seizures in patients. The idea behind the procedure was to reduce the spread of the epileptic neural discharges in the brain in order to control the severe seizures. Cutting the corpus callosum would not allow the abnormal discharge to spread throughout the entire brain. While the seizures were controlled as a result of the surgery and the patients did not demonstrate any major issues at first, there were also some other effects on those who had the surgery. Most of the communication between the left and right cerebral hemispheres of the brain was severely impaired. This led to the designation of so-called *split-brain patients*.

> You might often hear people saying they are "left brained" or "right brained" due to their talents or learning preferences. It is important to remember that the brain does not know the distinction between being "left brained" or "right brained." A person's strong points and their weak points result from complex interactions between genetics and environmental experiences. The distinction of being "left brained" or "right brained" is just a myth.

Research on split-brain patients has revealed many interesting findings. For instance, when an object was shown to the left visual field of these patients (seen only by the right hemisphere), subjects could not name the object later (a language function requiring the use of the left hemisphere) but could point to it with their left hand (a right hemisphere function because the right hemisphere controls the left side of the body). Some of these patients exhibited some peculiar difficulties, such as dressing themselves and then finding that one hand, typically their nondominant hand, would start undressing them. While this may sound humorous, these types of behaviors were quite annoying for these individuals. Appendix B includes a resource describing much of the research performed on these patients, the research findings, as well as left-brain/right-brain differences.

Cortical and Subcortical Areas

The cerebral hemispheres are covered with the *cerebral cortex*, which is a layer of tissue approximately three millimeters thick. As previously discussed, the cortex is convoluted (the *gyri* and *sulci*), but it is also made up of six parallel layers (called *laminae*). The laminae vary in thickness and prominence from one area of the cortex to another area, and a particular layer may be absent in certain areas of the cortex. Neurons in the cortex are not randomly arranged but are arranged in an orderly fashion. Some layers send projecting axons to other brain regions, whereas some layers receive projections from other brain areas. When brain tissue is preserved, the cortex takes on a grayish appearance, hence the term "gray matter."

Almost all the cortex one sees when viewing the brain is *neocortex* (new cortex). It is the neocortex that separates mammals from lower animals. Other forms of cortex are:

1. The mesocortex, consisting of the *cingulate gyrus* in the medial (middle) portion of the brain, between the hemispheres and above the corpus callosum, and the *insula*, which are cortical areas folded within the sulcus between the temporal and frontal lobes

2. The allocortex, which consists of the primary olfactory cortex and a structure called the hippocampus that is important in memory

Organization

The cells in the cortex are organized into columns of neurons with similar properties. These columns are perpendicular to the laminae (layers of the brain). The neurons within a given column have similar related properties and make connections with each other. For example, if one neuron in a column responds to touch on the right palm of the hand, then the other cells in that column will also respond to touch on the right palm.

Below the cortex is an area of white matter, which consists of the myelinated axons of neurons (hence the term "white matter"), and the area beneath this is referred to as the subcortex. The area beneath this is referred to as the *subcortex*. This includes a number of important brain structures that will be discussed in this book, such as the basal ganglia, thalamus, hypothalamus, brain stem, ventricles, etc. Many of the functions that occur in the subcortical areas of the brain are fast or reflexive and often are not mediated by conscious thought.

There are more than fifty areas in the cerebral cortex that can be identified based on their cellular differences. However, all of these areas have been grouped into four sections or *lobes*: the frontal, temporal, parietal, and occipital lobes of the brain.

Lobes of Fun

Dividing the brain into four lobes allows neuroanatomists to specify sections of the brain that are associated with a set of general functional capabilities. There are four different brain lobes; however, since there are left and right hemispheres, there are actually a total of eight lobes (four on the left side of the brain and four on the right side of the brain). **Figure** 2-3 illustrates these different lobes.

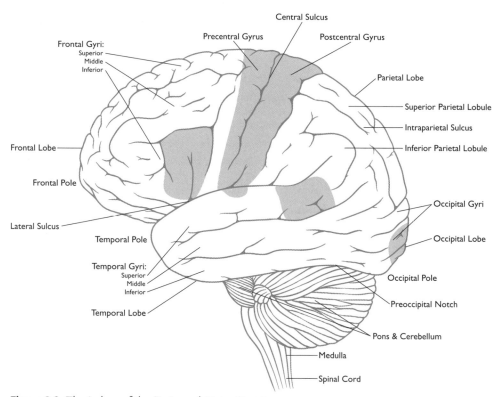

Figure 2-3: The Lobes of the Brain and Major Structures

While each lobe does seem to be tied to a set of specific functions, the brain is, in fact, a singular organ that performs many different functions in unison. In the same way that one can identify a person's arms and legs as different body structures, they are still parts of the body, a whole organism that acts in unison. Arms and legs do not function independently of the body. Likewise, the frontal lobe of the brain cannot perform any functions on its own. The brain cannot do anything without all of its parts; brain lobes do not function independently. Neuroimaging studies demonstrate that during any given action multiple brain areas are activated simultaneously.

The Frontal Lobe

The frontal lobe extends from the very anterior portion of brain to the central sulcus. The frontal lobe is considered to be the most developed portion of the brain in humans and represents relatively recent evolutionary changes in human brain structure compared to the brains of other animals. The most anterior portion of the frontal lobe is known as the *prefrontal cortex*. The prefrontal cortex receives information from all sensory systems and is primarily involved in functions such as planning, working memory (short-term memory), making abstractions, and other important functions that identify human thinking. The prefrontal cortex was the target of a surgical procedure known as a *frontal lobotomy* that disconnects the frontal lobes from the rest of the brain. This procedure was used on unruly children, people with depression, and people with other emotional problems or psychiatric disorders in order to correct these problems. There were over 20,000 lobotomies performed in the United States before the procedure was basically discontinued due to the development of psychiatric medications.

Two generally accepted subdivisions of the frontal cortex are the orbitofrontal cortex and the dorsolateral cortex. The *orbitofrontal cortex* is that part of the frontal cortex directly above the orbits or eyes. Based on studies of brain-damaged individuals, it appears that the orbitofrontal cortex is involved in the ability to inhibit emotionally inappropriate or socially inappropriate behaviors. The *dorsolateral cortex* is located toward the upper sides of the frontal cortex and appears to be associated with planning, ordering, and sequencing events.

> Sometimes lobotomies involved placing a flat metal instrument in the eye socket of an individual and moving it back and forth to disconnect the tissue (without any anesthetic, as there are no pain receptors in the brain). People who received frontal lobotomies generally lost their initiative, became somewhat impulsive, and had difficulties with certain aspects of memory and with emotional control.

At the very posterior portion of the frontal lobe lies the motor cortex, an area of the brain associated with voluntary movement. Further down the frontal lobe is an area known as Broca's area, which is associated with expressive language.

The Temporal Lobe

The temporal lobe is pictured in **Figure 2-3**. The temporal lobe is primarily involved in hearing, language comprehension, memory, and learning. The temporal lobe is often subdivided into four separate areas:

- Lateral (toward the side)
- Medial (toward the middle)
- Posterior (toward the rear)
- Polar region (the rounded portion)

The temporal lobe plays an important function in hearing as it contains the primary auditory cortex as well as the auditory association areas. A specific area in the temporal lobe, known as Wernicke's area, is associated with language comprehension. The medial portion of the temporal lobe is associated with the ability to learn and retain new information. An important brain area known as the hippocampus is located in the temporal lobe and is crucial for learning new information.

The Parietal Lobe

The parietal lobe lies behind the frontal lobe and is superior to (above) the occipital lobe, which is located at the back of the brain. The parietal lobe contains the primary association area and association cortices for the sense of touch (somatosensation). The parietal lobe also appears important in a number of other functions having to do with spatial information, such as imagery, rotating objects in your mind, and working memory. The posterior portion of the parietal cortex also appears important for storing representations of movements and controlling your intention to move.

The Occipital Lobe

The occipital lobe is found at the back of the brain and forms a boundary between the parietal and temporal lobes. The occipital lobe is extremely important in vision. This area of the brain contains the primary visual cortex and visual association areas that allow you to make sense of the information that your eyes transmit to the brain.

The Cerebellum

At the very posterior portion of the brain is a structure known as the cerebellum, see **Figure 2-3**. The cerebellum serves a number of important functions and has connections with nearly every other portion of the brain. It was originally believed that the cerebellum functioned only in body movements; however, recent discoveries have indicated that it is also important in certain aspects of cognition (thinking). The cerebellum has numerous connections with the frontal lobes and other areas of the brain.

Cover My Brain Please: The Meninges

The CNS contains very sensitive organs and therefore needs to be well protected. As discussed earlier, the spinal cord is surrounded by a long bony structure (the vertebrae) and the brain is encased in bone (the skull). The brain and spinal cord are also covered by three layers of protective membranes between the boney coverings and the actual CNS organ. These membranes vary in thickness and are collectively referred to as the *meninges* (the Greek plural for "membrane"). Working from the skull inward, they are the *dura mater, arachnoid*, and the *pia mater.*

The toughest layer of the meninges is the dura mater, which covers the inside of the skull and goes around the very tip of the spinal cord. At the bottom of the spinal cord, the dura mater forms a sac called the dural sac. At this location a physician might take a lumbar puncture, which consists of extracting a small amount of cerebral spinal fluid (CSF) to help identify certain infections, diseases, or other conditions. The CSF is found between the meninges and the bones covering the entire CNS. Neither the brain nor the spinal cord touch these bones and are suspended in this fluid, thus cushioning and protecting them. The CSF is also found in the brain ventricles (spaces) and helps protect the brain from blows to the head. Suspending the CNS in fluid also partially protects it from being bruised; however, in severe trauma, such as a severe head injury, the CSF may not protect the brain.

What is meningitis?

Meningitis is an infection or inflammation of the meninges and can be quite serious. Symptoms of meningitis include headache and neck stiffness, fever, confusion, vomiting, and becoming sensitive to light or noise (photophobia or phonophobia). A lumbar puncture can diagnose or exclude meningitis.

Holes in Your Brain: The Ventricles

There are four spaces in the brain known as *ventricles* that protect the brain from extreme pressure, which can be caused by head trauma. Each hemisphere of the brain contains one of the two lateral ventricles, the largest ventricles. Toward the posterior portion of the brain, the lateral ventricles connect to the third ventricle. The third ventricle connects to the fourth ventricle in the *medulla oblongata*, the portion of the brain that connects to the spinal cord. The ventricles contain cerebral spinal fluid (CSF), which is formed by groups of cells called *choroid plexus* located inside the ventricles. Since there is no "give" in the skull or vertebrae, if the brain or spinal cord were to be exposed to a condition that caused it to swell, it could become damaged quite easily. Among other things, the ventricles allow for some minor room for swelling. However, this allowance is very minor. The ventricles of the brain are also a communicating network of cavities filled with CSF.

The CSF flows from the lateral ventricles to the third ventricle and into the fourth ventricle. From the fourth ventricle some CSF flows into the central canal of the spinal cord (the central canal is a small hollow tube in the middle of the spinal cord that allows the CSF to flow in it); however, most of the CSF flows through an opening into the thin spaces between the brain and the meninges and then back into the ventricles (the spinal cord exits the skull through an opening called for *foramen magnum*). There is a constant process of absorption and production of CSF. In addition to providing a cushion for the CNS, the CSF supports the weight of the brain and provides hormones and nutrition for the CNS.

> When the flow of CSF is obstructed, it can accumulate, increasing pressure on the brain. In infants this may cause the skull bones to spread, causing an overgrown head. This condition is known as **hydrocephalus** and is usually associated with mental retardation in infants. Hydrocephalus can also occur in adults and can be associated with a number of physical and cognitive problems such as dementia.

Important Brain Structures

Several brain structures are so important that they deserve special discussion before continuing further.

The Thalamus

The thalamus is a complex structure and is the gateway to the neocortex. Any sensory information that projects from the body, with the exception of olfaction (smell), goes through the thalamus, is processed there, and then is sent on to the appropriate area of the brain. Sometimes information is sent back to the thalamus for further processing. The thalamus is also important in making fast links between multiple brain areas.

The Amygdala

The amygdala has extensive connections to many areas of the brain including sensory systems that project through the thalamus. The amygdala functions as a memory or association system for events that are emotionally salient. However, the amygdala does not produce conscious memories but instead produces unconscious responses to stimuli or stimuli that are associated with relevant stimuli. These responses are felt rather than thought, such as when you have a bad intuition or feeling about a person or situation but cannot explain why.

The Hippocampus

The hippocampus is part of a neural circuit that is involved in forming new memories and is found within the temporal lobe. The hippocampus does not store memories for long periods of time but is involved in the transfer of memory into long-term storage and has extensive connections with nearly every area of the brain. The hippocampus and amygdala are part of a larger network of structures called the *limbic system*, which is involved in a number of memory-related processes.

The Basal Ganglia

The basal ganglia are a complex set of subcortical structures that control movement and aspects of cognition. These structures have become well known due to their connection with Parkinson's disease.

The Cerebellum

The cerebellum has more neurons than any other structure of the brain. The cerebellum functions to coordinate motor behavior by analyzing programs for movement derived in the frontal cortex and comparing these to what is actually executed. The cerebellum also functions in learning, especially learning for motor movements. The cerebellum is found at the posterior portion of the brain right beneath the occipital lobe and directly above the spinal cord.

The Blood-Brain Barrier

Nearly 100 years ago it was found that when a dye was injected into the bloodstream, the tissues of the body, except for those of the brain and spinal cord, would turn that color. The blood-brain barrier (BBB) was thus discovered. The BBB is semipermeable and the endothelial cells that line the inner wall of the capillaries here are fit tightly together so that many substances cannot pass out of the bloodstream. Some molecules, such as glucose molecules that are used by the brain for energy, are transported out of the blood by special methods. Additionally, some substances, such as certain drugs and viruses, can also cross the BBB.

CAN YOU BELIEVE YOUR EYES?

The Process of Vision

THE PROCESS OF VISION is the most studied of all the human sensory systems. You may think that your visual system has developed to replicate an accurate representation of the world around you: An upside-down image falls on the retina and the image is then sent to your brain. Actually, nothing could be further from the truth. If that happened, what you would see would be akin to the frames of a two-dimensional, slow-motion picture. Read on and see why for yourself.

The Principles of Sensory Organization

The visual system does more than just attempt to produce an accurate representation of the environment in that the reconstruction of visual stimuli by the brain can appear different than reality. The visual system in the brain takes a very small, inverted, two-dimensional image from the receptors in the eye, deconstructs it, and recreates a detailed three-dimensional representation of the environment. In some respects, the visual system creates an even better representation or different representation of the real world. Before delving into the process of vision, it is important to understand the general principles of sensation, perception, and sensory organization.

Sensation and Perception

Sensation is a process whereby an organism detects stimulation from the environment, whereas *perception* is a higher-order process that involves integrating, recognizing, and interpreting complete patterns of sensation.

Sensation refers to how the physical attributes of the environmental stimulus activate the sensory organs and receptors in the organism, whereas perception refers to a broader process that involves sensation, the context, and the neural and cognitive processes utilized to make sense of the stimuli.

> Everyone knows that owls can see in the dark, right? Actually, this is a falsehood. No animal can see in complete darkness. Some animals, such as owls and other animals that are primarily nocturnal, have special adaptations that allow them to see under conditions of extreme dimness, but no animal can see in complete darkness.

There are three basic steps in the process of sensation and perception of a stimulus. The first step is *reception*, whereby stimulus molecules attach to the receptors in the body. In the second step, *transduction*, sensory receptors convert the energy of a chemical reaction that takes place in the receptors into action potentials (electrical signals). The third step is *coding*, in which spatial and temporal patterns of nerve impulses represent the stimulus in the brain in a meaningful way.

Areas of Sensory Cortices

The sensory areas of the cerebral cortex are considered to consist of three different types: *the primary sensory cortex*, *the secondary sensory cortex/cortices*, and *the association cortex/ cortices*. The primary sensory cortex receives most of its input from the thalamic areas associated with that system (except for the sense of smell). Recall that the thalamus is a sensory relay station that organizes information and sends it to other areas of the brain.

The secondary sensory cortex receives most of its input from the primary sensory cortex of that same system or other areas of the secondary sensory cortex associated with that sensory system. The association cortex of the system is any area of the cortex that receives input from more than one sensory system (primary and secondary cortices only evaluate information from one sensory system). Most of the input received by the association cortex comes from the secondary sensory cortex of the particular sensory system in question.

Interactions of the Sensory Cortices

The interactions of the three types of sensory cortices are defined by three major principles: *hierarchical organization*, *functional segregation*, and *parallel processing*.

The brain performs many functions simultaneously.

- Sensory systems are characterized by a *hierarchical organization* in that a system has levels or specific ranks that operate in relation to one another in much the same way that the military has a hierarchical system with generals, majors, captains, etc. Sensory systems are organized in a hierarchy based on how specific and complex their functions are. Each level of the sensory hierarchy receives most of its input from lower levels and adds an additional level of analysis before sending it up the hierarchy.

- *Functional segregation* refers to the notion that each of the three levels of the sensory cortices (primary, secondary, and association cortices) is composed of distinct areas that specialize in different types of analytic processes.

- *Parallel processing* refers to the fact that there are multiple pathways analyzing, interpreting, and sending sensory information at the same time. Recent findings in cognitive neuroscience indicate that there are two different types of overall parallel sensory streams that comprise sensory systems: one stream is capable of influencing behavior without conscious awareness, whereas another stream influences behavior by employing consciousness (this is sometimes termed the *dual process model*).

The Process of Seeing

The eyes of animals are special sensory receptors that capture light and send it on for further processing. *Light* can be conceptualized as photons, which are particles of energy, and at the same time light is conceptualized as waves of energy (contemporary physics views light as both a particle and a wave). The human eye responds to certain wavelengths of light, and these particles of light activate the receptors in the eye. The eyes do not "see" anything. The eyes capture light, transfer it to specialized cells that decompose the light (stimulus) and send this deconstructed information to different areas in the brain where it is reconstructed (perception).

Vertebrates have two eyes to see on each side of their body. Those with eyes on the front of their heads are generally predators, whereas those with eyes on the sides of their heads are prey animals. Front-facing eyes allow distance judgments, whereas side-facing eyes provide larger visual fields. Primates have front-facing eyes to navigate in trees where depth perception is very important.

Relevant Properties of Light

Wavelength and *intensity* are two properties of light that are important to understand in any discussion of vision. Wavelength of light plays an important part in the perception of color; the intensity of light in the perception of brightness. Wavelength is typically measured in nanometers (nm; billionths of a meter). Smaller wavelengths of light appear blue or violet, and higher wavelengths of light appear reddish. The perception of the light may not be the same thing as the light itself. For example, wavelengths of 700 nm appear to be bright red to us, but that does not mean that the light itself is red. The human eye will typically respond to wavelengths between 390 nm and 750 nm. Some animals can detect wavelengths that are much longer or shorter than can humans, such as rattlesnakes, which are able to detect infrared light waves that humans cannot see. **Figure 3-1** presents a schematic of the human eye.

The Eye

The iris—the band of contractile tissue that also gives eyes their color—regulates the amount of light that reaches the eye. Light enters the eye through the pupil, which is nothing more than a hole in the iris. The pupil size can adjust in response to changes in illumination (the amount of available light and the environment) so that the iris will

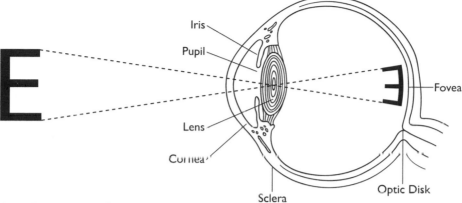

Figure 3-1: Diagram of the Eye

contract when light is very intense (bright) and expand in dim conditions. Adjustments in the size of the pupil also lead to changes in *sensitivity* (the ability to see under dim conditions) and *acuity* (detecting detail). When illumination is high, the pupils are constricted and the images that fall on the retina are sharper; however, under low levels of illumination the pupils dilate, letting in more light but sacrificing the focus of the images.

Behind each pupil is a lens that focuses the incoming light onto receptor cells within the retina, a light-sensitive membrane lining the back of the eyeball. Ciliary muscles are attached to ligaments that are attached to the lens. When you direct your gaze at something near to you, tension in these ligaments is adjusted so the lens can assume a cylindrical shape, allowing the lens to bend (refract) light and bring close objects into focus. When you try to focus on an object farther away, the ligaments are stretched and the lens becomes more flattened. This adjustment of the lens to focus light on the retina is referred to as *accommodation*.

> Myopia (nearsightedness) is better vision up close than at distances. The eye is typically too large, or the lens loses its flexibility and the image lands in front of the retina rather than on the retina. Hyperopia (farsightedness) is better distance than up-close vision and occurs because the eye is too small or because the lens becomes inflexible and the focused image falls behind the retina.

The movements of the eyes are coordinated so that every available point in the environment is projected to corresponding points on both the retinas. To accomplish this, the eyes converge, or turn slightly inward. *Convergence* of the eyes increases as you focus on things close to you, but the two eyes can never view the world from exactly the same position. This means that the difference in the position of the image on the retinas is greatest for objects that are very close. This difference, called *binocular disparity*, allows for the construction of a three-dimensional perception from an image on the retina that is two-dimensional. The reader can compare the disparate views of each eye by looking at the same scene and closing one eye at a time.

The Transduction of Light Into Neural Signals

Light passing through the pupil in the lens falls on the retina. The retina is a complex structure composed of five different types of neurons. Each of these five types of retinal neurons can be broken down into a number of different subtypes. Currently over fifty subtypes of retinal neurons have been classified. Figure 3-2 is a basic depiction of how these cells are layered.

The first thing one should notice is that the retina appears to be organized in a backward formation where light reaches the receptor layers after it passes through the four other layers. Once the receptors have been activated, the message is transmitted back through the retinal layers to the retinal ganglion cells, whose axons gather in a bundle and exit to the optic nerve, which sends the message to the cortex.

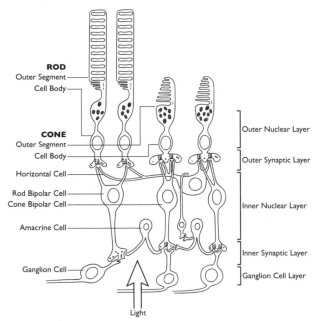

Figure 3-2: Retina of the Eye

> **Completion** is important in the function of the visual system. When a person looks at an object, the visual system does not conduct an exact image of the object to the cortex; instead it extracts the key features of the object, sends that information to the cortex, and the cortex reconstructs it, creating a perception of an entire object based on pieces of information.

This seemingly backward arrangement means that the incoming light is distorted by the retinal tissue it passes through before reaching the receptors. This problem is minimized by an area called the *fovea*, an indentation near the center of the retina that is specialized for high-detailed vision. The thinning of the ganglion cell layer at the fovea reduces the distortion of incoming light. What this also means is that the neurons in the retina do not transmit the image that falls on the retina to the brain.

Also related to the "backward" construction of the retinal tissues is the fact that the axons of the retinal ganglion cells that leave the eye are bundled together and exit to the optic nerve, creating a gap in the receptive layer called the blind spot (though there are still receptors here). Yet you do not see a hole in your visual field where these axons leave the retina. The visual system uses information provided by the receptors that are around the blind spot to fill in these gaps in your retinal image. This process is termed *completion* and is an example of how the visual system goes beyond creating an exact copy of the external environment. Thus, your brain can sometimes fool you by completing an image based on stimuli around the blind spot (as anyone who has driven a car and did not see another motorist behind them can attest to).

Retinal Layers

Of the five retinal layers of cells, the *amacrine cells* and *horizontal cells* are specialized to communicate across the channels of sensory input. *Retinal receptor cells, bipolar cells,* and *retinal ganglion cells* transmit their information directly to each other. The retinal neurons use both synapses and gap junctions to communicate.

There are two types of receptors in the retina of humans that are named after their basic shapes: rods and cones. Photopic vision refers to visual processes mediated by the cones, whereas scotopic vision refers to rod-mediated vision. The receptor cells (rods and cones) send their messages to the bipolar cells, which in turn send messages to the retinal ganglion cells. The retinal ganglion cells send messages to the brain. The function of the amacrine cells and the horizontal cells in the retina is to integrate the messages being sent from the retina. Horizontal cells also assist with vision under different light intensities; amacrine cells are inhibitory and also integrate information from the bipolar cells.

Cones are associated with vision that occurs in good lighting (photopic vision) and are also important in the perception of color. When light is dim and there is not enough energy to stimulate the cones on a reliable basis, the more sensitive rods predominate. Rods are specialized for vision in dimly lit areas (scotopic vision). These two types of receptors are connected differently with other retinal cells.

A small number of cones link to a bipolar cell and then to a retinal ganglion cell (a low-convergence system), whereas several hundred rods will link to a single retinal ganglion cell via several bipolar cells (a high-convergence system). A bipolar cell links with either rods or cones, but not with both. This results in dim light stimulating many rods that in turn influence the firing of the retinal ganglion cells; however, the same amount of dim light will not activate the cones to the same degree, and the retinal ganglion cells connected to cones will not fire in dim light.

Rods and cones are also distributed differently on the retina. There are no rods at all in the fovea, only cones. At the boundaries of the fovea, the number of cones begins to fall sharply. The rod density reaches its maximum at about twenty degrees from the center of the fovea.

Receptive Fields

The entire area of the world seen at any particular time is termed the *visual field*; the part of the visual field you see on your left is the left visual field and vice versa. Neurons also have *receptive fields*, which are the part of the visual field that any single neuron will respond to. The receptive field for receptors in the retina refers to the point in the environment where the stimulation activates the receptor. Neuroscientists can determine receptive fields in vision by shining light in various locations and recording the action of a particular neuron. When the image from the environment falls on the retina, it is decomposed by the layers of cells in the retina and sent to the brain.

Retinal ganglion cells send the messages from the other cells in the retina to the brain via the optic nerve. Retinal ganglion cells are designated as either X, Y, or W cells. W cells have small cell bodies and make up about 10 percent of the ganglion cells; Y cells have large cell bodies and also make up about 10 percent of ganglion cells; X cells have medium-sized cell bodies and make up about 80 percent of retinal cells. Y cells respond to large moving objects and are involved in the analysis of moving objects and directing attention to them. X cells appear to respond to smaller targets and analyze detail, high-resolution, and color. Such an arrangement is an example of parallel processing, where one particular cell type responds to movement, another to size, another to color, and so on—all of this occurring simultaneously. Most of this information is sent to the lateral geniculate nucleus (LGN) in the thalamus. The thalamic layers then project to the primary visual cortex.

> The lateral geniculate nucleus (LGN) is found in the posterior portion of the thalamus. The LGN is composed of six layers of neurons separated by axons and dendrites. Each layer receives its information from only one eye.

The Primary Visual Cortex

The primary visual cortex is located in the occipital lobe of the brain and is commonly referred to as the *striate cortex*. It was given this name because it contains a very characteristic stripe of white matter. Like most of the cortex, the striate cortex has six layers.

There are many pathways in the brain that are associated with visual information. The most thoroughly studied visual pathways are the *retina-geniculate-striate pathways*, which are the pathways that transmit signals from each retina to the primary visual cortex via the thalamus. (Recall that the thalamus is like a relay station in the brain that is responsible for coordinating and integrating information before sending it on to other areas of the brain.) These pathways are responsible for transmitting about 90 percent of the information from retinal ganglion cells.

> Some other axons go to an area called the **superior colliculus** or **tectum**, which is just below the thalamus and is involved in directing visual attention, and a few others travel to several other brain areas including a section of the hypothalamus that is involved in the sleep-wake cycle.

The optic nerves that transmit the information cross in the brain such that the signals from the left visual field are processed in the right primary visual cortex (the point in the brain where the optic nerves cross over is called the *optic chiasm*). All of the signals from the left visual field are transmitted to the right hemisphere—even information from the left visual field that is detected by the left eye.

What Was That Shape?

The neurons in the striate cortex have complicated receptive fields. A cortical neuron may respond to complex patterns as opposed to being excited or inhibited by light. The primary visual cortex is organized so that the flow of information goes from neurons with simple receptive fields to those with more complex receptive fields.

Most of the other neurons in the other layers of the striate cortex are composed of *simple* or *complex* cells. Simple cells, the smaller of the two, have receptive fields that can be divided into on and off regions and are also *monocular* (relating to one eye). The receptive fields of simple cells are rectangular rather than circular. These cells appear to be excited by parallel and/or perpendicular lines as opposed to light. Each simple cell responds best when its preferred straight-edge stimulus is in a particular position.

Complex cells also have rectangular receptive fields and respond best to straight-line stimuli in a specific orientation or to movement. Complex cells respond to stimulation from either eye (therefore they are *binocular*). However, these cells often display some degree of ocular dominance so that they respond more strongly to stimulation in one eye than they do to stimulation of the other eye. Other cells may fire best when the stimulus is presented to both eyes at the same time. These cells are likely to play an important role in the ability to perceive depth.

The neurons in the primary visual cortex are arranged in functional vertical columns with each cell in the column having a receptive area in the same area of the visual field. All the cells in the column respond to the same type of stimuli. The functional columns analyze input from one area of the retina.

The striate cortex is also organized so that different areas perform different functions. Thus, people with cortical deficits in one type of visual function may not have deficits in other functions. Studies indicate that perception of certain attributes occurs more quickly than perception of others. Color is perceived before motion. Color, facial processing, orientation, and other facets are perceived at different rates. However, depth, orientation, and left-right motion appear to be perceived at the same time.

> The flow goes from neurons with simple receptive fields to those with more complex receptive fields (hierarchical). Full images are *not* projected from the retina to the brain. The image falling on the retina is deconstructed and sent to different brain areas, reconstructed, and perceived.

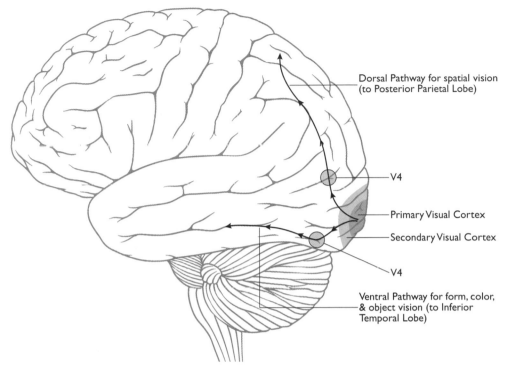

Dorsal Pathway for spatial vision
(to Posterior Parietal Lobe)

V4

Primary Visual Cortex

Secondary Visual Cortex

V4

Ventral Pathway for form, color,
& object vision (to Inferior
Temporal Lobe)

Figure 3-3: Visual Association Cortices and Where and What Pathways

Figure 3-3 depicts the primary and secondary visual cortices and two pathways: one for location and one for motion.

The Superior Colliculus

The superior colliculus is composed of several layers like the LGN and cortex, and receives visual input from the retinal ganglion cells. The superficial layers of the superior colliculus appear to receive input from visual areas (its innermost layers appear to receive input from other sensory systems such as auditory and somatic sensory systems). The primary function of the superior colliculus is the orienting of the head and eyes toward a visual stimulus. This is accomplished in unity with the frontal eye fields of the frontal cortex, which receive visual information from X cells in the visual cortex. Y cells appear to work through the superior colliculus in identifying visual outlines and with visual attention to objects. The superior colliculus may also control *saccades*, or the rapid movements of the eye.

One thing to understand from all of this is that visual system neurons are typically continually active even when there is no visual input, but they are specialized to respond to particular types of stimulation. Ongoing spontaneous activity is characteristic of most of the neurons in the brain. Brain activity is ongoing and does not terminate in the absence of environmental stimuli. Environmental stimuli either excite or inhibit neuronal activity.

How You See Motion

The majority of the research indicates that different parts of the brain are involved in the identification of objects and the location of objects. Visual information is transmitted from the lateral geniculate nuclei of the thalamus to the primary visual cortex. The information from the left and right lateral geniculate nuclei is sent to the striate, combined, and then segregated into different pathways that project separately to secondary visual cortices and then to association areas.

The specialized areas (secondary and association cortices) are believed to be parts of two major streams of information. A *dorsal stream*—(toward the top of the brain) which projects from the striate to the posterior parietal lobe—is known as the "where stream" and processes the object's spatial location and the direction of any movement it is taking relative to the viewer. A *ventral stream*—(bottom side of the brain) which projects from the striate cortex to the lower part of the temporal lobe—is known as the "what stream" and identifies and classifies the object.

Akinetopsia is a disorder in which the person cannot see movement progress in a smooth fashion. This disorder appears to be associated with damage to the middle temporal area—an area involved in the perception of motion, where the majority of neurons respond to specific directions of movement in the environment. Neuroimaging studies have indicated that this area is activated when someone views movement; blocking activity in this area produces motion blindness, and electrical stimulation of this area produces a perception of movement.

Much of the information concerning these two different streams comes from subjects who have brain damage in these areas and demonstrate selective impairments.

How You Perceive Color

The perception of color depends on the wavelengths of light that enter the eye. A number of theories have been postulated as to how the brain perceives color. *The trichromatic* or *component theory* proposes that there are three different cones that each have different sensitivities to wavelengths of light. The color of a particular object is a result of the ratio of wavelengths and activity in these three columns. The theory is based on the notion that any color can be created by mixing together three different wavelengths

The competing *opponent-process theory*, developed around the same time as the component theory (in the late 1800s), proposes that there are two types of cones for processing color and another receptor for processing brightness. Each of these three types of receptors is responsible for processing two complementary colors. For example, red and green are processed by one class of cells that change their activity in the direction of the wavelength. Other pairs would be blue-yellow and white-black. Several lines of evidence were used to support this viewpoint, one of which is the presence of afterimages that occur when staring at one color and then looking at a white background. If you stare at red long enough and then look at a white background, you will see a green shadow. As it turns out, components of both of these theories do appear to be relevant, but not fully explanatory.

Color certainly is perceived by the brightness and wavelength of light. More modern theories of color vision suggest that the color of something is determined by a proportion of light of differing wavelengths reflected off the surface of an object. The visual system then calculates this *reflectance*, compares the light reflected by surfaces adjacent to the object, and creates the perception of color. Thus, the perception of color depends on the analysis of the contrast in areas of the visual field and primarily occurs in the cortex.

How the brain separates energy and wavelength and then recombines them into color perception remains a mystery. Not everyone sees color. Color blindness can also occur via damage to an area in the cortical ventral stream producing a disorder known as *achromatopsia*. These individuals see only different shades of gray. Thus, while cones do contribute to the perception of color, the majority of the perception of color occurs in the cortex.

> The most common type of color blindness results from the absence of one cone type in the retina. The genes for these cones are coded on the X chromosome. It is more common for a male to have red or green color blindness as females have two X chromosomes and males just one (specific genes code for each cone). Blue-yellow color blindness occurs rarely.

Impaired Vision and Visual Illusions

The loss of sight is considered to be one of the most disabling things that can occur to a person. This section discusses various visual disorders and visual illusions.

Blindness

Most cases of blindness are due to retinal conditions, such as retinopathies and macular degeneration, which lead to the death of photoreceptors. Other conditions include glaucoma, which occurs most commonly due to excessive pressure within the eye and involves the death of retinal ganglion cells; cataracts, which cause clouding of the lenses; inherited metabolic disorders, such as diabetic neuropathy, which lead to the death of retinal cells; and eye and head injuries.

Visual Problems and Primary Visual Cortex Damage

Damage to a specific area in the primary visual cortex can lead to a *scotoma*, which is an area of blindness in the corresponding area of the contralateral visual field of both eyes. A perimetry test is used to determine these, where the patient's head is held motionless and he is instructed to stare at a computer screen, fixing his gaze on a central point. A small dot of light is flashed on various parts of the screen and the patient presses a button when he sees the light. This is repeated for both eyes and maps of blind areas can be developed. Because the brain engages in completion of blind spots, it is quite possible that individuals with massive scotomas are unaware of them. Because completion is performed by filling in holes in vision based on what the person sees, the person's perception can be inaccurate and lead to problems with driving, walking, and so on, until the person is formally tested.

Interestingly, many patients are unaware that they have cortical visual deficits. *Blindsight* is a phenomenon that occurs when patients respond to visual stimuli in scotomas even without conscious awareness of the stimuli. These individuals can see movement but claim that they cannot see objects in their visual fields. The perception of motion appears to be resistant to damage in the striate cortex. In some cases, it could be possible that some remaining vestiges of functioning striate cortex are still activated during movement and mediate vision in the absence of conscious awareness. The other explanation for blindsight is that there are visual pathways that project directly to the secondary visual cortex from subcortical areas without passing through the primary visual cortex. The evidence has never been conclusive for either explanation.

Can your brain ever lie to you?

Yes. Your brain fills in missing information based on existing information, groups like things together, and interprets three-dimensional images from two-dimensional pictures. The blind spot in vision is an example of how your brain lies by completing visual fields. What you see in this area often is not what is actually there.

Prosopagnosia

The term *agnosia* refers to a failure of recognition not due to sensory or verbal intellectual impairment. A visual agnosia is a specific agnosia for visually based stimuli. People with various types of visual agnosias can see certain objects but are unable to identify them. These can be specific to a particular aspect of visual input, such as movement, color, or specific objects, and supports the idea that various areas of the visual cortex (especially the secondary visual cortex) are specialized for these particular functions. One particularly interesting visual agnosia is *prosopagnosia*, a specific difficulty in recognizing faces.

Individuals with prosopagnosia can usually recognize a face as being a face but cannot recognize whose face it is, even if it is someone very familiar to them. The diagnosis of prosopagnosia is most often associated with damage to the ventral stream in the boundary between the occipital and temporal lobes in an area known as the *fusiform gyrus*. This area has come to be known as the "fusiform face area." Careful evaluation of these patients actually reveals that their deficit is not specifically with faces. For example, they can recognize a face as being a face but they cannot recognize whose face it is. Likewise, when evaluated further most of these patients can recognize a car as a car, a dog as a dog, etc., but cannot recognize *their* car or *their* dog. The fusiform gyrus appears to respond selectively to classes of visual stimuli and not just to faces specifically.

Anton's Syndrome

Visual anosognosia (Anton's syndrome) is a rare condition of cortical blindness that results from massive injury to the visual association cortex or occipital lobe. Affected persons cannot process the information sent from the intact eye or connections to the damaged cortex, but often deny that they are blind. They often confabulate to explain their visual deficits with remarks such as, "There is not enough light to see," or make up what they think they should see. Without the input of the visual-association centers, these people lose the concept of sight and cannot acknowledge their impairment. Anton's syndrome occurs most often in people with massive bilateral occipital damage.

Visual Illusions

Visual illusions are a result of perceptual processes that "trick" you. A branch of psychological thought called *Gestalt psychology* has investigated many of these processes and provided principles that explain them. Some links are provided for interested readers in Appendix B.

HEAR NO EVIL, SPEAK NO EVIL:
Language and Your Brain

AUDITION AND LANGUAGE play an extremely important part in your life. These are essential senses/functions that are involved in every factor of daily life from communication to survival. Do your ears hear and does your mouth speak? Hopefully by now you are thinking differently about your senses. So the answer to the old Zen question: "If a tree falls in the forest and no one is there to hear it—does it make a noise?" should be easy to answer at this point. If the answer is not clear now, it will be after this chapter. Read on, Grasshopper.

How You Heard That

Three elderly gentlemen are out walking. The first one says, "Windy, isn't it?" The second one says, "No, it's Thursday!" The third one says, "So am I. Let's go get a beer."

Sound is an extremely important perception and is important in directing behavior. Sounds are in reality vibrations of the molecules in the air that stimulate the auditory system. A person with "normal hearing" will be able to detect a sound wave between 20 and 20,000 Hz (hertz, or cycles per second), but humans are typically most sensitive to vibrations between 1,000 and 4,000 Hz. The frequency of the vibration determines its pitch (high or low). Dogs can hear pitches as high as 45,000 Hz and some other animals can hear much higher frequencies. **Figure 4-1** displays the anatomy of the ear for you to follow during the discussion of how these vibrations are interpreted as sound by the brain.

Sound waves travel through the air, reach the outer ear, and then travel down the auditory canal where they stimulate the tympanic membrane (eardrum). The signal is then transmitted by means of three small bones (*ossicles*) known as the malleus (hammer), incus (anvil), and stapes (stirrup) to the cochlea to other inner ear organs. The hammer vibrations lead to vibrations of the anvil and ultimately to vibrations of the

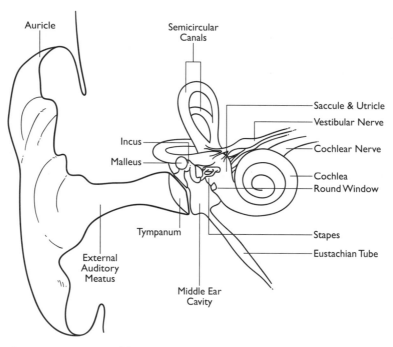

Figure 4-1: Diagram of the Ear

stirrup. The vibrations of the stapes then lead to vibrations of a membrane known as the *oval window*. The vibrations of the oval window transfer into the fluid contained in the snail-shaped *cochlea*. The fluid-filled cochlea is a coiled tube (hence the name cochlea, which means "land snail") that has an internal membrane running through it. The internal membrane is the auditory receptor organ known as the organ of Corti. The vibrations from the oval window move through the organ of Corti as a wave. This organ contains two membranes: the basilar membrane and the tectorial membrane. Hair cells, which act as auditory receptors, are situated in the basilar membrane, and the tectorial membrane rests on them (sort of like a hair sandwich). The vibrations from the oval window stimulate these hair cells, and this stimulation makes the axons in the auditory nerve fire (the auditory nerve is part of a branch of cranial nerve VIII). As the cochlear fluid vibrates, it is eventually dissipated by the round window, which is an elastic membrane of the cochlear wall. **Figure 4-1** also displays the semicircular canals, which are part of the *vestibular system*.

The human cochlea is extremely sensitive and can detect tones that differ in frequency by only two-tenths of a percent. Different frequencies produce their maximum stimulation at different points along the basilar membrane, with higher frequencies stimulating the area closer to the window and lower frequencies further up the membrane. This allows for a person to be quite sensitive to different tones and pitch.

How Sound Is Represented in the Brain

There is a network of auditory pathways from the inner ear to the auditory cortex located in the temporal lobe of the brain. The functions of the auditory cortex in humans are still not well understood, but projections from the cochlea lead to the *superior olivary nuclei* (or superior olives for the singular) located on both sides of the brain stem in the pons. These projections are *lateralized* such that sounds from the left ear are projected to the right superior olives and vice versa. Axons then project to the midbrain (to a structure called the *inferior colliculi*, where the information is integrated) and then to the thalamus (to a thalamic area called the *medial geniculate nucleus*). The information is analyzed in the thalamus, and if deemed important, it is then sent on to the primary auditory cortex, which is located in the superior temporal lobe. Routing sound impulses through two stops before going to the thalamus assists you in locating the source of the sound.

The primary auditory cortex (also called Heschl's gyrus) is slightly below the central sulcus in the upper-middle portion of the temporal lobe (see **Figure 4-2**). The neurons in the primary auditory cortex respond to different frequencies of sound depending on their location in the primary auditory cortex in much the same way that different neurons in different areas of the visual system respond to different shapes. Neuroanatomists refer to this frequency map as a *tonotopic* map, as different areas of the primary auditory cortex respond to specific tones. When one locates sounds coming from different locations, differences in the time of arrival of the signal from the two ears lead to differential stimulation in the superior olives, and this allows one to locate the direction of the sound.

Once the sound is processed by the auditory cortex, researchers believe it is sent via two streams of auditory analysis to two different areas of association cortices: an anterior auditory pathway located toward the frontal lobe that identifies the sound (the "what pathway") and a posterior auditory pathway more toward the parietal lobe that locates sounds (the "where pathway"), see **Figure 4-3**. Of course such an interpretation occurs at both of these levels of analysis for every sound. (Remember that any association cortex is an area of the cortex where sensory systems interact and associations are made. Areas of an association cortex have fields that react to visual information, fields that react to auditory information, and fields receptive to both visual and auditory information.)

Receptive Language

Slightly behind or posterior to the primary auditory cortex is an area known as Wernicke's area (see **Figure 4-2**). Wernicke's area is important in the comprehension of

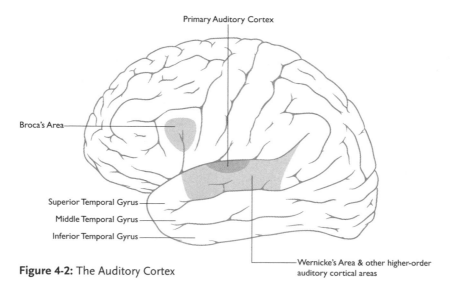

Figure 4-2: The Auditory Cortex

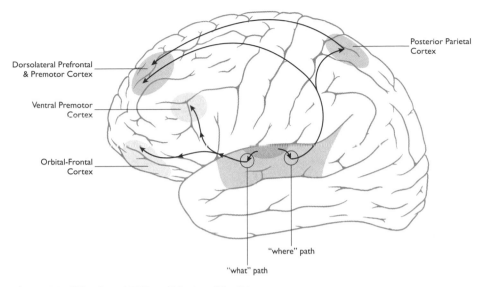

Posterior Parietal Cortex

Dorsolateral Prefrontal & Premotor Cortex

Ventral Premotor Cortex

Orbital-Frontal Cortex

"where" path

"what" path

Figure 4-3: "What" and "Where" Paths of Audition

language, both written and spoken. People with damage to this area develop difficulties understanding verbal language, a condition known as a receptive aphasia or Wernicke's aphasia (aphasia is a term used for language disorders).

Wernicke's area has extensive connections with another language area in the frontal lobe known as Broca's area, located in the frontal lobe just anterior (in front of) the primary motor strip that also resides in the frontal lobe (see **Figure 4-2**). Broca's area is important in the production of speech. Both Broca's and Wernicke's areas are located on the left hemisphere in nearly all right-handed people and most left-handed people (see the following for a discussion of how handedness relates to language). There is some evidence to suggest that the area on the right hemisphere that corresponds to Wernicke's area is important in processing prosody, the changes in tone and rhythm of speech that convey meaning.

Deafness

The common types of deafness are *conductive deafness*, associated with damage to the ossicles, and *nerve deafness*, associated with damage to the cochlea or to the auditory nerve. It appears that the major cause of nerve deafness is a loss of hair cell receptors in the basal membrane of the cochlea. If only certain hair cell receptors are damaged, the person may be deaf to some frequencies but not to others, something typically seen in age-related hearing loss. Typically this type of hearing loss occurs at higher frequencies. Some people with nerve deafness can benefit from cochlear implants that convert sounds picked up by a microphone on the patient's ear to electrical signals that are then carried to the cochlea. The signals then excite the auditory nerve. Deafness that results from damage to the ossicles is typically treated with a hearing aid that magnifies the amplitude of the sound waves.

> *Tinnitus*, or a chronic ringing in the ears, can result from a number of different causes, such as infection, exposure to loud noise, side effects from certain drugs, or even an allergic reaction. Tinnitus can also occur as a result of aging when people begin to lose their ability to hear at high frequencies.

The Process of Speaking

It is certainly true that animals can communicate with each other by making verbal calls. However, human language is not simply a type of communication similar to what other animals use when they make such calls. Human language differs from the verbalizations of animals on a number of different levels, the most important of which is a structured grammar that allows for complex ideas and abstractions to be communicated. Human language consists of rules of grammar, nouns, verbs, modifiers, etc., that allow for the transmission of ideas in a nearly infinite number of ways compared to the finite number of single words available for use. Attempts to teach primates language indicate that their messages are most often random with respect to the order of the words and that they are not capable of following or using the rules of grammar.

For most people, language functions are dependent on structures in the left hemisphere of the brain. Language researchers still do not understand why this is the case or why most people are right-handed, although there are many theories as to why this is so. Man's closest evolutionary relative, the chimpanzee, also demonstrates a particular hand preference; however, unlike humans, this preference appears to be a 50-50 split between left and right hands.

> Despite what is depicted in science-fiction movies, primates can only express what they have learned that a word is for, such as signing for food, or at best primates can express a very limited number of novel ideas consisting of two words. Primates do not learn sign language from humans and then sign to one another.

Wernicke's area of the brain—discussed earlier in this chapter—is important in the decoding of receptive language. A classic theory describing the brain structures involved in expressive language, receptive language, and reading is the *Wernicke-Geschwind Model*. According to this model, a person having a conversation will experience activation of the primary auditory cortex when hearing another person speak. This leads to activation of Wernicke's area, where spoken language (and written language) is processed. When one wishes to respond to the other person, Wernicke's area generates the neural representation of the thought or reply and sends it to Broca's area by way of a nerve tract known as the *arcuate fasciculus* that connects Wernicke's area to Broca's area. Broca's area activates the appropriate articulation for the response and sends it to the neurons that drive the appropriate muscles of speech in the primary motor cortex, the area of the brain that controls movement. Then the person responds.

When a person is reading, the signal received by the primary visual cortex is transmitted to the left angular gyrus in the parietal lobe. Here the visual form of the word is translated into its auditory code and then transmitted to Wernicke's area for comprehension. If you were reading aloud, this information would be transmitted back to Broca's area, and the previously described process would continue.

The Wernicke-Geschwind Model of language is based on studies of patients who had cortical damage and resulting language deficits. For example, Broca's area is named after the physician who discovered it, based on two patients with a severe expressive aphasia that resulted from damage to a specific area of the brain (the particular form of aphasia associated with damage to this area is often termed Broca's aphasia). Likewise, Wernicke's area is named after the physician who studied patients with a severe receptive aphasia who had damage to that particular area (the resulting aphasia is sometimes termed Wernicke's aphasia). However, it is rare to find patients that have circumscribed damage to one particular area as a result of a stroke or traumatic brain injury (in fact, recent reevaluations of both Broca's and Wernicke's seminal patients indicated that the brain damage was extensive in both cases and not circumscribed to those particular areas). Moreover, newer technologies such as functional brain imaging have indicated three important observations regarding language in humans that shed doubt on the Wernicke-Geschwind Model:

1. Language expression and language reception are mediated by many areas of the brain that also participate in the cognition (thinking) that is involved in language-related behavior.

2. The areas of the brain involved in language are not solely dedicated to language production or reception.

3. Because many of the areas of the brain that perform language functions are also parts of other systems, these areas are likely to be small, specialized, and extensively distributed throughout the brain.

Nonetheless, patients with left frontal brain damage near or around Broca's area will tend to demonstrate expressive language problems, whereas those who sustain damage around or near Wernicke's area are more likely to display difficulties with receptive language. Yet expressive or receptive aphasias are also seen in patients with damage to other areas of the brain.

> A disruption of language abilities due to brain insult is known as *aphasia*. Problems with articulation of speech due to nerve damage that are not impairments in language abilities are termed *dysarthria*. Dysarthria results from an insult to the motor-component of speaking such as the facial muscles and is not due to damage to the language centers of the brain.

Language in the Brains of Left-Handed and Right-Handed People

The vast majority of right-handed people have language lateralized to the left side of the brain (studies indicate that approximately 95 percent of these people have left-brain language). However, the situation is not quite the same with left-handed people. Left-handed people tend to be more ambidextrous than right-handed people as a group. The incidence of right hemisphere–dominant language in left-handers appears to be related to the degree of left-handedness in the person; the more left-dominant the person is, the more likely he will demonstrate a right hemisphere dominance for language. There are no significant differences in the normal language abilities of people with right-brained language. One study found the incidence of right hemisphere language dominance to be about 4 percent in strong right-handers (people who are predominantly right-handed), 15 percent in ambidextrous people, and 27 percent in strong left-handers.

How You Read

Reading is a complex process that occurs in the brain and may involve two different pathways: a *word-meaning-sound pathway* and a *word-sound-meaning pathway*. This model is referred to as the *dual-pathway* of reading.

The word-meaning-sound pathway involves recognizing a written word pattern (e.g., the word "dog") as being the same pattern as one stored in memory. If the written word pattern is linked to a particular meaning in the association cortex, one becomes conscious of the meaning of the word. This pathway is alternatively referred to as the *whole-word* route. This route takes some time to develop because when one is first exposed to words, one has not made the connection between the word pattern and the meaning of the word. However, once these connections are formed, they work efficiently and quickly.

Reading using the word-sound-meaning pathway involves breaking up the written word pattern into segments (e.g., "d-o-g") and combining the segments to form a mental representation of the sound of the word. The sound of the word is then compared to sounds in the storehouse of word sounds in the brain and recognized as the same sound as the one in memory. Because the stored word sound is linked to a meaning, you then become aware of the meaning. Some refer to this pathway as the *sound-of-letters* pathway.

> ***Modular models*** claim language comprehension is executed in independent brain modules working in a one-way direction so that higher levels cannot influence lower levels that have completed their functions. ***Interactive models*** state that all types of information contribute to word recognition, including the context; words are recognized as the result of nodes in a network that are activated together. ***Hybrid models*** attempt to combine both.

Supporters of the dual-pathway model of reading believe that in most adults the two pathways complement each other and work in parallel, allowing for the quickest and most accurate reaction. This model also explains mistakes people sometimes make when reading, such as when a person speaks a different word from what is written, but the word still conveys the same meaning. For example, while reading out loud one might say "they went to sleep" while reading the phrase "they went to bed." Such errors indicate that the person is extracting the meaning before decoding the word, thus using the word-meaning-sound pathway.

Dyslexia

The term *dyslexia* refers to a group of disorders that involves impaired reading. There are two types of dyslexia: *developmental dyslexia*, where a person has difficulty reading despite normal intelligence, normal development, and no history of brain damage (true dyslexia); and *acquired dyslexia*, which occurs in an individual who previously had no significant difficulty reading but due to some type of brain injury has developed problems reading (sometimes called *alexia*).

Research on dyslexia indicates that there are probably a number of different brain mechanisms that, when disrupted, result in dyslexic behaviors. For example, one view of developmental dyslexia is that the dyslexic person cannot process incoming visual information. This disruption is believed to occur in one of the thalamic visual system pathways, the magnocellular pathway. However, other researchers suggest that this pathway is not impaired in all dyslexics and that the cerebellum, the structure in the posterior portion of the brain, is impaired. Based on neuroimaging studies of individuals with dyslexia, these researchers suggest that this dysfunction in the cerebellum interferes with the performance of many automatic and over-learned skills (which the cerebellum moderates), and reading is one of these. Imaging studies have sometimes shown that some dyslexics demonstrate abnormalities in the cerebellum and not in a thalamic pathway.

The Effect of Brain Damage on Language

The effects of damage to the auditory cortex have been difficult to determine because of the location of the auditory cortex. Most of the human auditory cortex is located in the lateral fissure of the brain. It is rarely entirely destroyed, but there is often extensive damage to the surrounding tissue. Studies of the effects of damage to one side of the auditory cortex indicate this results in difficulty localizing sounds coming into the opposite ear (e.g., left auditory cortex damage would result in difficulty localizing sounds coming from the right side).

A small number of cases with damage to both sides of the auditory cortex have been studied, and initially there appeared to be a total loss of hearing in these cases; however, hearing in these cases often recovered weeks later. It may be that the total hearing loss occurred as a result of the initial shock of the damage, and as the neurons recovered hearing was restored. Over the long run, the major effects of bilateral auditory cortex damage appear to disrupt the ability to localize sounds and the ability to discriminate between different frequencies.

As discussed earlier, aphasia can be associated with brain damage that can occur in many different parts of the brain. The production of speech in aphasia is classified as either fluent or nonfluent.

Four language-related factors are generally considered when assessing aphasia: expressive language (length of phrase, articulation, and other factors), language comprehension (usually for simple or complex commands), naming ability (typically naming pictures of objects), and repetition (the ability to repeat single words or phrases).

Fluent aphasias are characterized by verbal output that is plentiful, well-articulated, and of relatively normal lengths and prosody (prosody refers to variations in pitch, rhythm, and loudness). Nonfluent aphasias are characterized by verbal output that has short phrase lengths, is effortful, and has disrupted prosody. Fluent aphasias are typically associated with more posterior brain damage (e.g., Wernicke's aphasia), whereas nonfluent aphasias are associated with more anterior brain damage in the motor regions of the brain that involve speech production.

Nonfluent Aphasias

- *Broca's aphasia* typically occurs when there is brain damage to the posterior portion of the frontal lobe of the brain. Verbal output is described as telegraphic in that it is pressured, slow, and halting. Articulation is impaired and utterances are typically less than four words in length. These patients can repeat single words and very short phrases but have difficulty naming objects. Comprehension of speech is relatively good except for complex commands or passive sentences.

- *Transcortical motor aphasia* involves lesions or other areas of damage in the frontal lobe (but not at Broca's area). People with this type of aphasia make little attempt to spontaneously speak, and when they do speak, their utterances are typically less than four words, but articulation may be good. Comprehension is generally good; naming objects is good; and repetition is generally preserved.

- *Mixed transcortical aphasia* results from diffusion lesions surrounding cortical language areas. Expressive language is severely limited or absent altogether; naming is significantly impaired; and comprehension is impaired. Repetition, despite impaired comprehension of language, remains generally preserved.

- *Global aphasia* results from massive damage to the language-dominant hemisphere and results in markedly impaired expressive and receptive language abilities.

Aphasia is not typically an isolated consequence of brain damage. The specific areas of the brain that contribute to language production and reception are rarely damaged in isolation. Often there are serious motor deficits, such as paralysis to the opposite side of the body, sensory deficits, neurological deficits, and other cognitive deficits, associated with these aphasias.

Fluent Aphasias

- *Wernicke's aphasia* occurs with lesions or damage to the posterior portion of the temporal lobe. Expressive language is fluent with normal utterance length but with an abundance of semantic and phonemic *paraphasias*, such as word substitutions or neologisms (made-up words). Because these patients cannot comprehend what they say or mean to say, their expressive language, while fluent, is often meaningless and incomprehensible (often referred to as "word salad" due to many words being spoken, but without any coherence). Comprehension is impaired; naming is severely disrupted; and repetition is typically severely impaired.

- *Conduction aphasia* occurs with damage to the supramarginal gyrus in the parietal lobe or to the arcuate fasciculus. Conversational speech is relatively normal in length but marred with paraphasias. Comprehension is relatively spared, but the person may have difficulty with complex verbal statements or following multistep commands. Repetition is poor and naming is always impaired.

- *Anomic aphasia* can occur when there is damage in the angular gyrus. Conversational speech is typically good but with word-finding difficulties. Repetition is generally good, and auditory comprehension is typically good, except during instances where the person must divide her attention between tasks. Naming is impaired in these patients in light of an absence of other significant language problems.

- *Transcortical sensory aphasia* most often involves damage at the junction between the temporal, parietal, and occipital lobes of the brain. Conversational speech is fluent but with word-finding difficulties and semantic paraphasias (word substitutions that are often totally inappropriate or do not resemble the intended word in sound or meaning). Comprehension is typically impaired, but repetition is surprisingly good, in light of poor comprehension. Naming is typically impaired.

DO YOU FEEL ME?

The Sense of Touch

If you were to place your hand on the speaker of a radio or stereo system while it is on, you could feel the vibrations the sounds make. These vibrations are what you "hear." However, if you lost your hearing, you could never train yourself to experience the vibrations in your hand as sound. Nonetheless, the human body does have a well-developed sense of touch. This chapter discusses the *mechanical* senses (other than hearing): touch (pain, temperature changes) and balance (the ability to ascertain limb and body position).

How You Obtain Balance: Vestibular Sensation

Feedback from the visual system (which acts as a motion detector) and the vestibular system is responsible for maintaining your balance. The *vestibular system* is located in the inner ear and consists of the three semicircular canals that are next to the cochlea (see **Figure 4-1**). Like the organ of Corti in the cochlea, the semicircular canals are filled with fluid and contain hair cells that detect the fluid's movement. Imagine a pole going through the top of your head to the floor (a vertical axis) and a horizontal sheet through your head from ear to ear (a horizontal axis). The semicircular canals detect rotation along these axes as well as any movement along the horizontal axis from front to back.

Signals from the semicircular canals are involved in the *vestibulospinal reflex*, a balance reflex that involves the cortex, the cerebellum (the structure at the back of the brain), and the spinal cord. Signals from the semicircular canals are combined with signals from the retina and cortex that detect motion in the same direction detected in the semicircular canals. This allows for the visual system and the vestibular system to work in unison to obtain your orientation in space, to feel and coordinate motion, and to help with balance control (mostly for balance standing up). This system for balance control operates through your core or trunk. When you fall or slip in one direction, say toward the left, you may extend your left leg and left arm to counteract the fall and restore your balance. This system signals for that corrective response.

> A person with vestibular damage would have trouble reading a street sign while walking. This is because the vestibular system allows the brain to shift eye movements to compensate for changes in head position. Without this ability, the experience of reading while walking would be like trying to read a severely jiggling book.

The sense of balance is accomplished by complex neural processes that are distributed throughout the spinal cord, vestibular system, visual system, and the cerebellum. Of course, your sense of balance is strongly associated and works with the ability to move. However, some researchers consider the sense of balance a sixth sense that should be included with the other five senses and insist that balance is a "perception" much like vision or hearing.

Touch Me: Somatosensation

Sensations from the body are called *somatosensations*. The somatosensory system is made up of three interacting systems:

1. An *exteroceptive* system that responds to stimuli applied to the skin (there are three subdivisions here: touch, temperature, and pain receptors).

2. The *proprioceptive* system that monitors information about the position of the body (this includes the vestibular system previously discussed).

3. The *interoceptive* system that provides information concerning conditions inside the body, such as temperature.

The exteroceptive system is the focus of this section. The skin has a number of different receptors. The simplest cutaneous (skin) receptors are called *free nerve endings*, which are unspecialized neural endings that are particularly sensitive to temperature changes on the skin and to painful stimuli. The *lamellated corpuscles* are the deepest and largest cutaneous receptors. These receptors respond to sudden changes in pressure on the skin (they respond quickly), whereas *Merkel's discs* and *Ruffini endings* respond to gradual pressure and skin stretch (both of these receptors respond more slowly). **Figure 5-1** shows a cross-section of the skin.

When the skin is touched, or pressure is applied to the skin, the stimulation results in a firing of all the receptors, which leads to the perception of being touched. After a very short time (a few hundred milliseconds), the slowly adapting receptors remain activated and the fast receptors deactivate, which changes the quality of the perception.

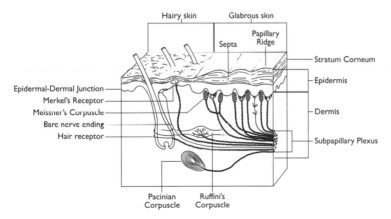

Figure 5-1: Receptors in the Skin

Imagine putting on a hat: at first you can feel it on your head, but rather quickly the perception goes away unless you focus on it. Receptors that adapt at different rates allow for the organism to receive information concerning static and dynamic qualities of stimuli. Somatosensory receptors are specialized in such a manner that a particular type of receptor responds to a particular type of stimulus; however, the receptors all function in a similar manner. When they are stimulated, the chemistry of the receptors is altered, allowing the cell membrane to exchange ions in the same manner as occurs during the action potential of a neuron. The neural fibers that carry information from the somatosensory receptors (skin and other receptors) join together as nerves and enter the spinal cord by way of the *dorsal roots* (on the back of the spine).

Dermatomes are areas of the body that are innervated by a particular segment of a dorsal root spinal cord section. If a specific dorsal root nerve is destroyed, there is often not significant somatosensory loss as there is considerable overlap between dermatomes that are next to each other. A reference for viewing the dermatomes is provided in Appendix B.

Skin to the Brain

There are two major pathways that transmit somatosensory information from the body to the brain. The *dorsal column–medial lemniscus* system tends to relay information about touch and proprioception, whereas the *anterolateral* system tends to relay information about pain and temperature changes to the brain. However, the disconnection of one system does not eliminate the type of information sent to the brain by that system, so the division of these functions between the two systems is not complete. Both systems must relay both types of information to some extent.

The dorsal column–medial lemniscus sensory roots (for touch) enter the spinal cord from their sources by means of a dorsal root (a path in the back of the spinal cord) and travel up in the dorsal columns to the *dorsal column nuclei* in the medulla (where the spinal cord connects to the hindbrain). It is at this point where they synapse (relay the signal). The axons of the dorsal column nuclei cross over (decussate) to the other side of the brain and connect to the *ventral posterior nucleus* of the thalamus by way of nerves known as the *medial lemniscus*. The ventral posterior nucleus also receives projections from three branches of the trigeminal nerve that carry information from the opposite side of the face. Most of the neurons in the ventral posterior nucleus of the thalamus send projections to the primary somatosensory cortex of the brain; however some send projections to the secondary somatosensory cortex in the posterior parietal lobe.

Are all neurons the same length?

Neurons can be of differing lengths depending on their targets. The longest neurons in the human body are the dorsal column neurons that begin in the toes and run up to the medulla.

In contrast to the dorsal column–medial lemniscus, the neurons in the *anterolateral dorsal roots* (pain perception) synapse when they enter the spinal cord. The axons of the second order neurons either decussate (cross over) and continue up to the brain or move up the spinal cord on the same side as they entered. There are three different tracts of the anterolateral dorsal root system:

1. A *spinothalamic tract* that projects directly to the ventral posterior nucleus of the thalamus

2. A *spinoreticular tract* that projects to the reticular formation and then to the thalamus (at two different sites called the parafascicular and intralaminar nuclei)

3. The *spinotectal tract* that connects to the tectum (roof) of the midbrain

The three branches of the trigeminal nerve previously mentioned also carry information regarding pain and temperature from the other side of the face to these sites in the thalamus. Once the information reaches the thalamus, it is then sent on to the primary somatosensory cortex or the secondary somatosensory cortex.

The signals of touch and pain are processed in the thalamus and then sent to the appropriate cortical areas. The primary somatosensory cortex (sometimes referred to as SI) is located directly posterior to the central fissure at the very anterior portion of the parietal lobe and is organized according to a map of the body surface commonly called the "somatosensory homunculus" (*homunculus* means "little man"). Thus, the somatosensory cortex is often referred to as being somatotopic in organization. **Figure 5-2** displays the organization of the primary somatosensory cortex.

The "homunculus" in the somatic sensory cortex is disproportional in that more area of the somatosensory cortex receives inputs from the parts of the body that make tactile discriminations, such as the hands, lips, tongue, etc., than from other parts of the

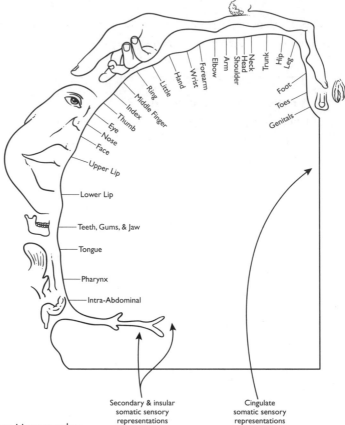

Figure 5-2: Sensory Homunculus

body. Areas of the body, such as the back, have proportionately smaller representations in the somatosensory cortex. The secondary somatosensory cortex (sometimes called SII) lies inferior to the primary somatosensory cortex, and much of it extends into the lateral fissure. The secondary somatosensory cortex receives a great deal of its input from the primary somatosensory cortex and thus receives information from both sides of the body, whereas the primary somatosensory cortex receives its input from the contralateral (opposite) side of the body (cortex on the left side of the brain receives information from the right side of the body and vice versa). Both areas of the somatosensory cortex send much of their output to the association cortex in the posterior portion of the parietal lobe of the brain. Neurons in the primary somatosensory cortex are either excitatory or inhibitory and display the columnar organization discussed in Chapter 3. Neurons in a column tend to respond to the same type of stimulation, whereas neurons across columns respond to different types of stimulation.

The signals from the somatosensory cortex are sent to areas of the prefrontal cortex and posterior parietal cortex association areas. The association areas in the posterior parietal cortex contain neurons that respond to both somatosensory stimuli and visual stimuli (neurons that respond to two types of stimulation are called *bimodal neurons*). The two receptive fields of each neuron are related in a spatial manner; for example, if a particular neuron's somatosensory receptive field is in the right hand, then its visual field is next to the right hand. As the hand moves, the visual receptive field moves with it.

Different Distributions

One of the reasons for the discrepancy in the amount of different cortical areas allotted to different areas on the body is that different areas of the skin have different densities of receptors for touch. The fingertips and tongue have many more receptors than do similarly sized areas of skin located on the back. These areas of higher receptor density allow for much finer discriminations.

Anticipation

There is evidence that parts of the cortex can become more or less activated when one anticipates that they are about to be touched. For example, one study found that activity in the primary sensory motor cortex increased, and activation in the areas outside the primary sensory motor cortex decreased, when people were anticipating being tickled, a pattern similar to activity observed during actual tickling. The motor area of the frontal cortex must receive projections from the somatosensory system in order to make a decision about what to do in response to stimulation. When it receives this information, the motor area and the frontal cortex can generate the appropriate action.

What a Pain: Pain Perception

Touch and pressure are signaled by the corpuscles in the skin, whereas pain is signaled by the *nociceptors*, which are specialized cells that may be myelinated or unmyelinated. The myelinated fibers conduct information about pain very quickly and, when these cells are activated, there is usually an immediate reaction, such as when you touch a hot stove and pull your hand away immediately even before you're aware of what you've done. The unmyelinated cells are involved in the duller and longer-lasting types of pain.

The perception of pain is actually considered paradoxical by many brain researchers. The perception of pain actually has no apparent representation in the human cortex. In fact, the perception of painful stimuli appears to result in the activation of many areas of the human cortex, but these areas vary from study to study and between participants in studies. Painful stimuli typically lead to activation in the primary and secondary somato-sensory cortices; however, these brain areas do not appear to be necessary for one to experience pain. For instance, studies of individuals who have had a cerebral hemisphere surgically removed indicate that they can still experience pain on the side opposite the removed cerebral hemisphere. Pathways for pain to the CNS are diffuse, have short-term or long-term effects, and can have many different connections.

While painful experiences are indeed unpleasant, another paradox of pain is that these experiences are often necessary and important for survival. Pain is designed to be a sort of intense warning to stop some harmful or potentially harmful activity or to seek treatment for an injury. However, an interesting paradox of pain is that it can often be suppressed by emotional or cognitive factors. Athletes have been known to accomplish amazing feats with broken bones that would make most people fall down and cry. Some religious ceremonies require participants to pierce their bodies, and yet these participants often express or feel little pain, and soldiers and people injured in life-threatening situations frequently feel no pain until the event is over.

A theory regarding pain that has been quite influential is the *gate control theory* of pain. This theory proposes that signals descending from the brain are able to activate neural mechanisms in the spinal cord (neural gates) that can block incoming neural pain signals to the brain. If a person is sufficiently distracted, the gate can close and the experi-ence of pain can be lessened or blocked altogether.

Pain Control

There is evidence for the gate control theory of pain. The cerebral aqueduct is in the midbrain, contains cerebrospinal fluid, and connects the third and fourth ventricles. The gray matter around the cerebral aqueduct is called the periaqueductal gray matter (PAG). The PAG contains special receptors for opiate-based drugs like morphine and codeine. Why would the brain have built-in receptors for synthetic drugs? As it turns out, your brain also produces several endorphins, which are endogenous opiate neurotransmitters that can block the perception of pain. The discovery of these endorphins and the PAG answered questions as to why people could block pain perception under certain circum-stances as well as why certain drugs relieve pain.

The brain has a built-in pain control system that occurs when output from the PAG excites a cluster of nuclei in the core of the medulla, known as the raphe nuclei, causing them to release the neurotransmitter serotonin. The serotonin release results in a signal sent on the dorsal columns of the spinal cord that blocks incoming pain signals into the dorsal horn of the spinal cord (recall that pain signals enter the dorsal part of the spinal cord from the skin and other parts of the body). It has also been hypothesized that the placebo effect that is often seen in studies where participants, who are given a pill that they are told will block their pain but are actually given an inert compound, still report a reduction in their pain. Somehow, the placebo interacts with the patient's beliefs and activates the gate blocking mechanism.

Pain perception can work both ways, however. For example, some of the descending neural circuits can increase, instead of blocking pain, as when someone is fearful or under a great deal of stress. Pain perception appears to activate the anterior cingulate cortex of the brain. The cingulate cortex is located in between hemispheres just above the corpus callosum (in the mesocortex). The front part of the cingulate cortex, the anterior cingulate cortex, may be a sensitive monitoring center in the brain. This area is activated by painful stimuli, the anticipation that one will experience a painful stimulus, and failure in obtaining one's goals. The purpose of such a center in the brain appears to be concerned with sorting out different strategies to obtain goals. For example, if you were bitten by a large brown dog, you may find yourself feeling anxious whenever you see a large brown dog, thus resulting in avoiding the dog and avoiding potential pain.

> Different stimuli are often paired together, as in the anticipation of pain and going to the dentist. Sometimes two different stimuli become associated with a particular outcome. This process is called *conditioning* and is a form of learning that can produce voluntary or involuntary responses to the stimuli.

Differences in Pain Perception

Pain thresholds and individuals are variable, and the ability to tolerate pain is also quite variable among different people and within the same person in different situations. Some studies suggest that pain tolerance increases as people get older, but it is not clear as to whether this finding is due to decreased sensitivity in the system, psychological factors, or other factors. Athletes and people with high levels of motivation appear to have higher pain tolerances, and there have been suggestions that people from different cultural backgrounds display different tolerances for pain. This last finding may be related to the cultural appropriateness of expressing pain that differs from culture to culture.

> **Peripheral neuropathy** occurs when pain receptors in the peripheral nervous system become inactive or die as a result of vascular problems. This leads to a loss of feeling in the affected limb. Peripheral neuropathy typically occurs in the fingers or toes of affected individuals.

The Phantom Limb Phenomenon

Sometimes your brain can fool you as in cases of people who have phantom limb sensation or phantom pain. The phantom limb sensation results from the ability of the brain to change in response to stimuli. This physical change in the brain tissue is known as *neuroplasticity* or just *plasticity*. Most individuals who have a limb amputated continue to have the sensation that the limb is still present. This experience is termed a *phantom limb*. Even though the limb has been amputated, the neurons in the somatosensory cortex still continue to send signals as if the limb were still there. Eventually the phantom limb sensation will dissipate. Phantom limb sensations result from the neuroplasticity of the brain. Because there are no incoming signals from the amputated limb, the neurons associated with the phantom limb, needing stimulation, may connect with adjacent cortical areas. Often, stroking the area of the body whose cortical representation is adjacent to that of the amputated limb produces the sensation that the amputated limb is being stroked (e.g., stroking the left side of the face or left shoulder produces the sensation that the amputated left hand is being stroked).

> **Neuropathic pain** is chronic pain that occurs without a recognizable cause. This typically develops after an injury has healed, but the person continues to experience extreme pain. The mechanism producing neuropathic pain is unknown, but recent research suggests that abnormal glial cells in the brain produce signals that lead to hyperactivity in the pain pathways. Neuropathic pain can also occur following an amputated limb.

Phantom pain is experienced in nearly half of all amputees. The mechanism for this extreme pain is not yet understood, but it is hypothesized that either it is due to changes in the primary sensory cortex after amputation (neuroplasticity), conflicting signals received from the amputated limb and from the visual system that sends motor commands to the amputated limb, or innate memories of limb position.

CHEMICALS IN YOUR BRAIN:
Taste and Smell

THE FIRST SENSORY SYSTEM in animals was probably a chemical sensitivity of some type. Chemical senses are important in finding food, locating potential mates, and avoiding danger. Although most people would think that vision is the most important sense, in terms of survival value, chemical senses have far more to offer for most animals. This chapter reviews the chemical senses of taste and smell in humans.

How Many Different Tastes Are There?

Taste receptors are found on the tongue and in parts of the mouth (oral cavity). These receptors typically occur in groups of approximately fifty, termed *taste buds*. Taste buds located on the tongue are most often found in the *papillae*, which are the small protrusions located on the tongue. The average number of taste buds a person has is estimated to be at around 10,000. There is not a one-to-one correspondence between a taste bud and a neuron; the neurons responding to taste buds receive many inputs. You have probably heard at some time or another that there are four or five primary tastes. Typically these primary tastes consist of bitter, salty, sour, and sweet. In some cases meaty (*umami*) is included as a primary taste. Other newer models of taste include these five primary tastes plus *metallic* as a primary taste, making the number of basic tastes the tongue can detect as six. Other models extend the number further.

Can you taste water?

Dogs and pigs, and perhaps some other animals, can taste water, but people cannot. Humans do not actually taste water, even mineral waters. What humans taste are the chemicals or the impurities in the water.

While taste receptors interact with smell, this conventional view of taste purports that every taste that humans experience results from the combination of one or more of these primary tastes. One of the difficulties with this notion is that many tastes that people experience could not be formed by a combination of these five primary tastes. Nonetheless, this particular model continues to dominate the field regarding the experience of human taste as no empirically validated alternative models have yet been proposed.

The Brain and Taste

Taste signals leave the mouth via three cranial nerves:

1. Cranial nerve VII, the facial nerve, carries information from the front of the tongue.

2. Cranial nerve IX, the glossopharyngeal nerve, carries information from the back of the tongue.

3. The vagus nerve, cranial nerve X, carries information from the back of the oral cavity.

> Ageusia is the inability to taste, but it is rare because taste is transmitted by three neural pathways. Partial ageusia is more common, usually following ear damage on the same side of the head. The loss of taste is restricted to the anterior two-thirds of the tongue as cranial nerve VII carries taste information by passing through the middle ear before synapsing in the medulla.

All of these fibers travel to the medulla where they synapse on a structure called the *solitary nucleus*. These taste signals project to the thalamus. From the thalamus, neurons project to the primary gustatory cortex, which is near the section of the somatosensory cortex that represents the facial area. Some neurons from the thalamus also project to the secondary gustatory cortex, which is in the lateral fissure of the brain. The projections from the taste receptors in the tongue to the brain do not decussate (cross over) to the other side of the brain as projections for other senses do.

> A supertaster experiences a much greater sensitivity and intensity in the sense of taste than the average person due to having far more papillae than average. These people tend to need less sugar and fat in their foods; however, due to their heightened taste for bitterness, they prefer salty foods. Supertasters are more likely to be females of Asian, South American, or African heritage.

The sense of taste is modulated by two satiety mechanisms. The first of these occurs in the brain, and some texts refer to this as a central mechanism. After you have been eating for a while, neurons that respond to both smell and taste in the orbitofrontal portions of the brain begin to fire more slowly. As a result, you begin to lose your preference for the item that you are eating. This process is called *alliesthesia* (changed taste). A more quickly working mechanism occurs in both the taste buds and olfactory (smell) receptors called *sensory-specific satiety*. Appetite is also regulated by hormones from the gastrointestinal tract that affect the brain in the hypothalamus and brain stem.

That Smell Went Right to Your Brain

The receptor cells for smell (olfaction) are embedded in the olfactory mucosa, a layer of tissue covered with mucus in the upper part of the nose that also contains supporting cells and basal cells. The dendrites for the olfactory receptor cells are located in the nasal passages, and their axons pass through the *cribriform plate*, a porous portion of the skull. Unlike other senses, the neurons for olfaction do not synapse in the thalamus but instead enter the olfactory bulbs and synapse on neurons that project to the olfactory tracts located in the brain.

> Bears have one of the most highly developed senses of smell, over 2,000 times more acute than a human's. A bloodhound has a sense of smell about 300 times that of a human. However, snakes may have the advantage when it comes to smell, as they are able to collect molecules in the air with their tongues and transfer them back to their brain via an organ known as the Jacobson's organ.

Olfaction is a very important sense, and other mammals have larger areas of the brain devoted to it. It is believed that bypassing the thalamus results in a time-saving mechanism basic to survival, as scent is important in enemy detection and the mating of many animals. In humans, it is estimated that there are about 10 million olfactory receptors and approximately 1,000 different receptor types. At this time researchers have not been able to determine if there is some type of organized distribution of all of these various olfactory receptors in the olfactory mucosa; however, it appears that all of the receptors project to the same area in the olfactory bulb.

Olfactory bulbs are made up of six layers of different types of neurons. Olfactory receptor axons terminate near the surface of the olfactory bulbs in groups called *olfactory glomeruli*. Each of these gets its input from many olfactory receptor cells. It appears as if the glomeruli are arranged in a systematic fashion because their layout is similar in related species and there is a mirror symmetry between left and right olfactory bulbs (the glomeruli that respond to a particular odor are located on the same site of each olfactory bulb). Olfactory receptor cells demonstrate a constant process of deterioration and replacement. New olfactory cells develop axons that grow and extend to the appropriate part of the olfactory bulb. This process of deterioration and regeneration occurs every few weeks.

The smells that people experience are made up of a mix of many different odors that produce hundreds of signatures across the olfactory bulb. The olfactory bulb has several main projections that go directly to areas of the cortex without going to the thalamus first. This makes the sense of smell unique among all of the other senses.

The olfactory bulb projects directly to the orbitofrontal cortex and then relays to the mediodorsal thalamus, which sends projections back to the orbitofrontal cortex. This projection allows for conscious attention and awareness to odors. The interconnection with the thalamus interacts with other sensory information to integrate the sense of smell with visual, auditory, and somatosensory input. For instance, when standing at the seashore, you see the ocean waves coming in; you hear the sound of the waves crashing on the shore; you feel brisk salty air; and you smell the combination of salt-fresh air and seaweed. These all make a lasting impression on you.

Olfactory Projections

Projections to the orbitofrontal cortex pair smells from food with information from taste receptors to create perceptions of flavor. This is why food tastes bland when you have a cold. There are also approach/avoidance programs associated with this tract that work in unison with smell-related information from other tracts ("programs" are stored action patterns of behavior).

The olfactory bulb sends projections to an area of the cortex known as the piriform cortex (sometimes spelled pyriform cortex), an area of the medial (middle) temporal cortex. The piriform cortex is considered the primary olfactory cortex by many researchers. This area also appears to be involved in approach/avoidance behaviors associated with smells.

> Dementia is a brain disorder that typically occurs in elderly people and most often initially presents itself as a loss of memory. The connection between the hippocampus (a brain structure important in creating new memories) and olfaction is so strong that specific tests for dementia have been developed that measure the loss of smell in an affected patient.

The olfactory bulb also projects to the entorhinal cortex, an area in the medial (middle) temporal lobe that is important in memory. The entorhinal cortex interfaces with the hippocampus. This projection allows for the memory of odors that were important to you. The olfactory bulb also has projections with the amygdala, an area of the brain that is important in emotional responses. The amygdala, positioned in front of the hippocampus, specializes in important memories with emotional overtones. Because certain smells have inherent emotional aspects to them, the amygdala allows for emotional reactions to important smells. A large percentage of smells evoke emotional reactions, such as the smell of freshly baked bread or the smell of skunk. The amygdala also connects to the hippocampus so that there is a brain circuit that is involved with memories that have emotional overtones, such as the smell of rotten food or odors associated with sex. Projections from the amygdala also go to the hypothalamus, an area of the brain that triggers the release of certain hormones.

Pheromones

In humans, the main role of the chemical senses of taste and smell is recognition of flavors. However, in other species of animals chemical senses also play an important role in regulating social interactions between members of the same species. Many animals release pheromones, which are chemicals that can influence the behavior of animals within the same species. For example, when a female dog is "in heat," she releases pheromones that notify male dogs that she is primed for mating.

There has been some attention, especially in the business world, to the possibility that humans may also release pheromones. Some findings have supported this notion. For example, the menstrual cycles of women living together tend to become synchronized; the olfactory potential of women is at its most sensitive when they are ovulating or when they are pregnant; and some men can judge the stage of a woman's menstrual cycle based on her vaginal odor. However, despite the huge commercial market for human pheromones, there is no direct empirical evidence that human odors serve as sexual attractants. However, certain smells do directly affect the behavior of humans.

CHAPTER 7

———◆———

YOU MOVE ME

A BODY MOVEMENT as simple as reaching up to touch one's nose requires a complex combination of neural signals to be sent within the brain, down the spine, to the arm, and then to the hand. A number of potential programs are activated and inhibited by brain mechanisms, but only the relevant programs are followed. In this chapter the process of movement is discussed.

Muscles and Their Movements

Before discussing how the brain is involved in movements, it is important to understand some basic terms regarding how the voluntary muscles work. *Effectors* are muscles. *Flexors* are muscles that bend a joint, whereas *extensors* are muscles that straighten a joint. So the biceps muscle in the upper arm acts as a flexor and the triceps muscle acts as an extensor. If two muscles work together, they are *synergistic*, whereas those muscles that act in opposition are *antagonistic*.

Muscles are very elastic, can stretch (within limits of course), and are attached to the bones by tendons (ligaments attach the bones to one another). In most cases, muscles have to be told to move.

Motor Neurons, the Spinal Cord, and the Muscles

The major interplay between the muscles and the nervous system is mediated by *alpha motor neurons* that begin in the spinal cord, exit through the ventral root of the spinal cord (belly side), and project into the muscle fibers. An action potential in an alpha motor neuron releases the neurotransmitter acetylcholine in the muscle. The alpha motor neurons translate nerve signals into a mechanical action by changing the length and tension in the muscles. The release of acetylcholine results in the muscle fibers contracting. Alpha motor neurons get their input (instructions) from a number of sources.

> The primary neurotransmitter involved in movement at the muscle junction of the nerve is acetylcholine, but other neurotransmitters involved in body movement work in the brain, such as dopamine and GABA.

At the lowest level of the hierarchy, the alpha motor neurons receive input from sensory fibers in the muscles when stretched. This causes a sensory signal to be generated in the joints and muscles that is then transmitted to the dorsal roots of the spinal cord where the signal then synapses directly on the alpha motor neurons. If the muscle stretch was unintentional, this signal returns them to their normal length. This is the process of many reflexive actions such as the patellar reflex. At the level of the skeletal muscle, these activities are monitored by the *Golgi tendon organs* and the *muscle spindles*. The Golgi tendon organs respond to increases of tension in the muscle to provide feedback about muscle tension and prevent damage during extreme muscle contractions.

The muscle spindles are located in the belly of the muscle and detect changes in muscle length, giving the CNS important information about the position of the body. Moreover, muscle spindles are involved in the *stretch reflex*, which keeps external forces from altering the body position and counteracts the stretch of the muscle. The aforementioned patellar tendon reflex is a stretch reflex.

Myasthenia gravis is a disease that appears to be autoimmune in nature, although there is a congenital form. In this disease the receptors in the muscle for acetylcholine become dysfunctional, leading to weakness, fatigue, and in advanced cases to paralysis. Some medicines that inhibit the breakdown of acetylcholine can be effective for a short time, but the disease is progressive and eventually these medicines become ineffective.

Some muscles are antagonistic, meaning they work in the opposite direction from one another. This could pose a problem for smooth movement if it were not for the principle of *reciprocal innervation*. Reciprocal innervation refers to the notion that antagonistic muscles are innervated to allow a smooth motor response; when your biceps muscle contracts, your triceps muscle automatically relaxes.

At the upper end of the hierarchy, alpha motor neurons receive inputs from descending fibers from the brain traveling through the spinal cord and interneurons within the spinal cord segments. Descending motor fibers originate in several cortical and subcortical areas of the brain. The signals sent down them can be either inhibitory or excitatory and provide the message for voluntary movement. For instance, commands sent from the cortical areas of the brain can activate the arm so you can reach out and stretch your arm. As you reach out and stretch your arm, the resulting action in the alpha motor neuron to your triceps muscle is excitatory, whereas the signal to the biceps is inhibitory.

Units of Movement

Like the sensory systems, the *sensorimotor system* is organized in a hierarchical fashion from the brain through the spinal cord to the muscles of the body. This allows the components of the higher levels to be involved in the more complex aspects of movement and for the lower ranks (e.g., the muscles) to carry out the commands from higher levels. This hierarchy works in a parallel-processing manner such that the signals between the higher levels move over multiple paths at the same time. Such an organization allows the higher levels (e.g., the association cortex) to operate and manipulate the lower levels in a number of different ways.

> There are three types of muscle: Striated, or skeletal, muscle consists of long cylindrical fibers that are striped; smooth muscle is found in the intestines and other organs and consists of long thin cells; cardiac muscle is found in the heart and consists of fibers that are fused together at various points. Cardiac muscles contract together, not independently, as do skeletal muscles.

Each level of the hierarchy is often composed of different units that perform different functions. This type of organization is termed *functional segregation*. The movement accomplished by the sensorimotor system is guided by sensory feedback and sensory input. The entire system operates on the principle of neuroplasticity such that experiences change the nature and physical structure of the system. For example, when a person learns how to type, this produces changes in both the muscles and tendons involved in typing as well as in the brain areas that send and receive messages involved in the learning process. Movement is a complex function mediated by many systems of muscles, tendons, nerves, and neurons.

The muscle system contains many neural structures that begin in the brain stem. Two of the prominent subcortical structures involved in muscle movement are the *basal ganglia* and the *cerebellum*.

The Basal Ganglia

The basal ganglia (BG) are composed of five nuclei or structures. These include the caudate, the putamen, the globus pallidus, the subthalamic nucleus, and the substantia nigra. The BG are not a singular anatomic structure as much as they are a functional series of interconnected structures with limited inputs and outputs. The BG are quite complicated and almost all of the incoming projections to the basal ganglia are targeted at the caudate and the putamen, which together are known as the *striatum*. The incoming fibers can come from many areas of the cortex including motor, sensory, and cortical association areas. Most of the output from the BG originates in the globus pallidus and the substantia nigra. (The projections from the substantia nigra primarily synapse in the superior colliculus and are involved in the initiation of eye movements.)

Outputs from the globus pallidus terminate in the thalamus. The thalamus sends messages from these projections to the motor cortex, the supplementary motor area in the cortex, the prefrontal cortex, and the anterior cingulate gyrus of the limbic system. The BG send their outputs via two paths: a direct path that goes through the brain stem and an indirect path that projects to the thalamus and then to the premotor and primary motor cortices in the brain.

It appears that a major function of the BG is in the inhibition of movement. The basal ganglia allow for the motor system to be kept in check and for all possible cortical representations of movement to become activated without actually engaging the muscles in action. Once a specific muscle movement plan gains strength and is decided on, the inhibitory signal for that plan is decreased for those select neurons and the plan is implemented.

The Cerebellum

The cerebellum is a large structure that receives numerous inputs from the visual, auditory, vestibular, and somatosensory areas of the brain as well as many areas of the association cortex. The cerebellum contains more neurons than the rest of the entire cortex and is involved in maintaining balance, muscle movement planning, the execution of muscle movements, and in movements of the eyes. Like the basal ganglia, the cerebellum sends its outputs via two paths: a direct route that goes through the brain stem and an indirect route that projects to the thalamus and then to the cortical areas.

> Guillain-Barre syndrome is an autoimmune disorder that occurs when the immune system of the body attacks the nerves in the peripheral nervous system, resulting in the demyelination of the nerves outside the central nervous system. Nerve transmissions in the muscles are affected, and extreme muscle weakness and paralysis in the person occurs. Neurons in the brain and spinal cord are typically not affected.

How Your Cortex Moves You

At the head of the hierarchy for body movement are the association cortices. Unlike other sensory systems discussed so far, the sensorimotor system is primarily controlled by two areas of the association cortex: the posterior parietal association cortex and the dorsolateral prefrontal association cortex. There are other areas of the brain that join in as well.

The portion of posterior parietal cortex that relates to the primary somatosensory cortex is termed the *posterior parietal association cortex.* This area of association cortex is important in directing your behavior by providing spatial information (e.g., depth, distance, etc.) as well as directing your focus. This area is believed to receive information from the visual system, the auditory system, and the somatosensory system to allow you to localize the body and other objects in the environment. Much of the output from this area is then sent to the motor cortex (located at the posterior portion of the frontal lobe just anterior to the somatosensory cortex), the dorsolateral prefrontal association cortex, areas of secondary motor cortex, and to the frontal eye fields in the prefrontal cortex that control eye movements.

The *dorsolateral prefrontal association cortex* receives information from the posterior parietal cortex and then sends the information to the primary motor cortex, secondary motor cortex, frontal eye fields, and back to the posterior parietal cortex. The dorsolateral prefrontal association cortex is important in evaluating stimuli in the environment and initiating voluntary movements based on context. The neurons in this area of the cortex are specialized to evaluate the stimuli in the environment and initiate responses to them. Neurons respond to characteristics of stimuli, location of stimuli, or both location and object characteristics, whereas other neurons are stimulated by the actions or responses of

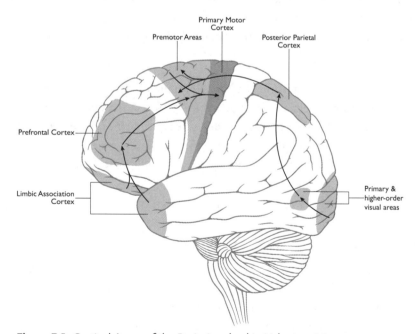

Figure 7-1: Cortical Areas of the Brain Involved in Voluntary Movement

environmental objects. Some neurons appear to be activated by one's intent to respond. **Figure 7-1** displays the posterior parietal association cortex and the dorsolateral prefrontal association cortex along with other cortical areas.

Areas of the secondary motor association cortex receive input from the primary association cortices and send much of their output to the primary motor cortex. There are multiple areas of secondary motor cortex and supplemental motor areas in the human brain. These secondary areas include the supplementary motor area, the premotor cortex, cingulate motor areas, and others. Areas of secondary motor cortex are connected with association, and other secondary motor areas are believed to be involved in specific movements or patterns of movements that are initiated by the dorsolateral prefrontal cortex. Neurons in the areas of secondary motor cortex appear to become activated prior to the initiation of a voluntary movement and remain activated throughout the completion of the movement. These are also pictured in **Figure 7-1**.

The *primary motor cortex* is located right on the precentral gyrus in the posterior portion of the frontal lobe. This motor strip, as it is often known, is directly anterior to the somatosensory cortex. It is this portion of the brain that is directly involved in voluntary movement, as all other areas—such as the premotor, association, supplementary, and subcortical motor area—send their information here. The motor strip then categorizes it and sends the command to the appropriate muscles.

In 1937, the neurosurgeon William Penfield and his associates performed a very famous series of experiments that mapped the primary motor cortex by applying low-level electrical stimulation to various points of the motor strip and observing which part of the body moved as a result of the stimulation. What these researchers found was that an area on the contralateral side of the body moved when a particular area of the motor strip was electrically stimulated. The resulting information indicated that there is a motor homunculus that is organized in a way so that larger areas of the cortex are devoted to controlling areas of the body associated with more intricate types of movements, such as the hands.

Patterns of Movement

More recent research, using longer bursts of brain stimulation at slightly higher intensities than those used by Penfield, resulted in complex sequences of motor responses instead of singular movements. There still appears to be a somatotopic organization in the primary motor cortex (e.g., stimulation to the hand area produces movements of the hand); however, the responses from these stimulations were complex movements often involving several different parts of the body as opposed to the individual muscle contractions that Penfield had noted. Sites that Penfield originally believed to be involved in a particular body movement were also found to overlap with other sites that moved other areas of the body. The research now suggests that signals from sites in the primary motor cortex diverge to allow a particular site to move a body part to its targeted location no matter what its starting position is. This means that the sensorimotor system must be quite flexible so that it can respond and change over time based on the needs of the organism.

> Like the sensory systems, the motor systems are hierarchical, guided by sensory (especially somatosensory) feedback, and are physically changed by the amount of prior practice/learning that occurs in the system. Lower portions of the hierarchy send their messages to higher areas. The primary motor cortex then sends the commands to the appropriate areas.

Descending Pathways

The instruction from the brain must be sent down the spinal cord and to the targeted area. Several different descending pathways carry signals from the cortex and brain stem to the alpha motor neurons in the spinal cord. The signals sent through these pathways act together in the control of voluntary movements. Descending motor tracts from the brain can originate in the cortical areas (these fibers are termed *pyramidal* or *corticospinal tracts*) or in the subcortical areas of the brain (these fibers are known collectively as *extrapyramidal tracts*).

> You place your hand on a hot stove; immediately sensory pain receptors send a signal to a sensory neuron in the spinal cord that synapses on an interneuron. This excites a motor neuron, sending the signal back to the muscles to pull away. Your cortex is not involved in these *reflex arcs* that are patterned early in development; otherwise you would suffer more physical damage.

Motor control is lateralized in the same way that sensory representation in the brain is lateralized. Each cerebral hemisphere is devoted to controlling movement on the opposite side of the body. The exception to this lateralization occurs with fibers projected from the cerebellum so that cerebellar functions are ipsilateral (the right cerebellum is involved in motor control of the right side and vice versa).

Programs in Your Head

It has been hypothesized that the whole sensorimotor system is made up of programs organized in a hierarchy such that the lower levels of the sensorimotor system have certain patterns programmed into them, and the higher levels produce complex movements by activating different combinations of these programs. This allows for the same task or movement to be carried out in a number of different ways. For instance, the aforementioned studies of primary motor cortex stimulation indicated that stimulation to certain areas resulted in patterns of behavior and not just singular movements. Investigation of subcortical structures also appears to support these hypotheses.

> It appears that shifts in brain activation occur during motor movements. Learning a new motor activity appears to be associated with activation in the lateral premotor and prefrontal cortical areas, whereas performing learned behaviors appears to be associated with activation in the supplementary motor area and in the hippocampus.

The basal ganglia are involved in movement and fine motor control, the initiation of movement, and especially the inhibition of movement. The basal ganglia have extensive connections with the prefrontal lobe of the brain. These structures are composed of phylogenetically older structures that are also found in lower vertebrates such as frogs. In humans, the basal ganglia appear to have influences in both simple movements and in the ability to engage in complex short-term and long-term goals. One of the main functions of these structures is to monitor all potential cortical motor plans and inhibit plans or movements except for the movement that is to be executed. For example, a person running from a bear will not suddenly stop and try to scratch his foot.

The basal ganglia perform other modulatory functions but send out few descending fibers to motor areas. The structures in the basal ganglia are part of the neural loops that receive input from various cortical areas and send it back to the cortex by way of the thalamus. A great number of these signals go to and from motor areas of the cortex. The basal ganglia also appear to perform a number of cognitive functions and are involved in the perception of reward and punishment.

The cerebellum receives multiple sensory inputs from peripheral receptors, association areas of the cortex, and the prefrontal cortex. The cerebellum has connections with the primary motor cortex and the prefrontal and frontal areas of the brain. The cerebellum consists of three different anatomical regions that have different sets of connections. Because of these numerous connections, the cerebellum receives information from primary and secondary motor cortices, signals from the brain stem motor neurons, and also feedback from vestibular and somatosensory systems.

The most posterior section of the cerebellum controls eye movements and projects to the vestibular system for the correction of balance and motor learning. The medial portion of the cerebellum has connections to the primary motor cortex, which then projects to the spinal cord to control the finer details of movement and correct errors in movement and goals. The more lateral portions of the cerebellum project to the thalamus and to premotor areas to assist in error correction regarding planned movements and future plans. For instance, if you are about to try to hit a baseball, there is information stored in the cerebellum regarding the sequence of hitting the ball, past experiences at batting, and correction programs for hitting the ball. Diffuse damage to the cerebellum results in a loss of the ability to control the direction, speed, and/or sequences of movement as well as difficulties with speech and eye movements.

Movement Disorders

Some movement disorders are a result of cortical brain damage. For those who suffer from them, they can have a significant impact on their lifestyle. The type of disorder and its effects depend on where and how much damage is incurred. A few movement disorders are listed here.

Hemiplegia

Damage to the cortical areas of the primary motor cortex often produces *hemiplegia*, which is paralysis on the opposite (contralateral) side of the body. Hemiplegia will often result from a stroke, particularly in the middle cerebral artery, which is a large artery that feeds the lateral and superior portions of the brain. Reflexes are typically absent immediately after a stroke, but may appear later and may be hyperactive due to the removal of inhibitory mechanisms. The degree of hemiplegia that occurs in a stroke or in brain damage depends on the area of the brain affected and/or the extent of the damage, and the degree of recovery is variable. Physical rehabilitation methods have been shown to be useful in helping a patient with hemiplegia recover some of her functioning; however, often there are residual effects even in patients with good recovery. Weakness occurring on one side of the body not associated with paralysis is termed *hemiparesis*. Numbness following a stroke often occurs in varying degrees with hemiplegia and hemiparesis or may occur independently of these disorders.

Apraxia

Damage to the cortical areas of the brain can lead to deficits with motor coordination. In apraxia, there is typically no hemiplegia but a loss of motor skills, which depending on the area of the brain lesion (damage), can be specific or general in nature. There are two major categories of apraxia.

- An *ideomotor apraxia* presents as a disturbance of voluntary movement such that the person cannot decode the intent or idea into movement. The person can recognize an object and a plan to use the object, but cannot voluntarily implement the plan. The connections between the processing and planning areas to the motor cortex are disrupted. These individuals can still perform automatic movements (like using a toothbrush), but they cannot perform them when requested to do so and cannot imitate them.

- *Ideational apraxias* occur when the person's knowledge of the action is disrupted. The patient has difficulty identifying the concept or purpose behind an object and cannot form or recall a plan of action for object use. Sequences of movements are either incomplete or completed in improper sequence. The processing and planning areas are damaged, typically in the parietal lobe in these apraxias.

Nearly all patients that have a type of apraxia also typically have some language difficulties (aphasia), but the opposite situation does not seem to be the case. Thus, apraxias are often associated with left hemisphere brain damage as for most people language is a left hemisphere function.

Subcortical Damage

There are several disorders that occur with damage to the basal ganglia. The two major disorders are:

PARKINSON'S DISEASE

Parkinson's disease (PD) is a progressive neuromuscular disorder that has an average age of onset in the fifties. In the vast majority of cases, there is no genetic association. The cause of PD is unknown, but the neuropathology is relatively well understood. The neural loops connecting the basal ganglia and the thalamus are either a direct route that promotes actions (excitatory) or an indirect (inhibitory) route that inhibits motor action.

Two pathological characteristics of PD are: a loss of dopamine-producing neurons and the appearance of Lewy bodies.

The loss of dopaminergic neurons is the characteristic feature of PD. The principal pathology of PD affects the dopamine-producing neurons in the substantia nigra of the basal ganglia, where most of the dopamine in the brain is synthesized. There is a massive loss of these neurons in this area and dopamine becomes depleted.

The other characteristic feature of PD is the appearance of Lewy bodies, which are abnormal clusters of protein that develop in nerve cells. The accumulation of high levels of Lewy bodies leads to neuronal degeneration and to cell death in the brain.

The loss of dopaminergic neurons in the basal ganglia leads to the classic triad of symptoms found in PD: tremor, muscle rigidness, and bradykinesia (slowed movements). Tremors in PD include a fine motor tremor, usually most pronounced at rest (resting tremor) but exacerbated with movement. Bradykinesia is also evident in the general lack of excessive movements, such as decreased expressive gestures, decreased facial movements, and a lack of facial expressions. Rigidity occurs in the joints and in the shuffling gait that occurs in PD patients, where patients walk as if their feet are magnetized and stuck to a metal floor.

The movement disorder in PD can be sporadic as there is another subcortical route that is involved in gross motor movements that bypasses the basal ganglia via the cerebellum. So PD patients are sometimes able to walk or run normally in emergency situations. Treatment for PD often consists of dopamine agonists (medicines that enhance dopamine actions) and Levadopa, which replaces dopamine. About 40 percent of PD patients develop dementia.

HUNTINGTON'S DISEASE

Huntington's disease (HD) is a hereditary disorder tied to a dysfunction in a gene for a protein called huntingtin that leads to an unstable condition in the basal ganglia. The disease causes degeneration in many regions of the brain and spinal cord, especially in the subcortical brain regions. Symptoms of HD usually begin when patients are in their thirties or forties, with the initial symptom being depression or other personality changes such as irritability or anxiety. There is a progression to severe dementia, with many individuals exhibiting psychotic behaviors. The main symptom in HD is *chorea,* involuntary, jerky, and sometimes violent arrhythmic movements of the body. These can be accompanied by gait disturbances, muscle weakness, and clumsiness. The average life expectancy after being diagnosed with HD is about fifteen years.

Cerebellar Disorders

Because the cerebellum has three functional divisions, damage to the cerebellum can result in a number of difficulties. The cerebellum controls eye movements, fine motor control, balance, arm movements, and others. Cerebellar damage can lead to *asynergia*, a loss of coordination of motor movement; *dysmetria*, which is an inability to judge distance and when to stop; *adiadochokinesia,* difficulty performing or an inability to perform rapid alternating movements; *intention tremors*, which occur during movement but not at rest; balance difficulties: *gait ataxia,* wide-based walking and staggering; *hypotonia,* muscle weakness; *nystagmus*, abnormal eye pursuit movements and *dysarthria*, slurred speech.

Cerebellar agenesis is a rare inherited disorder in which the person is born with a partial formation or total absence of the cerebellum. Those born with a partial formation may display few or no symptoms. Total absence often results in poor muscle tone, tremors, uncontrollable eye movements, and/or poor muscle coordination.

HOT, HUNGRY, AND THIRSTY:
Internal Body States

SOME OF THE MOST MOTIVATING factors in life have to do with maintaining balance in your internal body states. However, while the body attempts to maintain a comfortable internal state, external conditions can affect this balance. Unlike other body states, hunger and thirst are subject to environmental control and cannot be fully controlled by internal mechanisms. No one likes to be hungry or thirsty, or too cold or too hot. This section will briefly look at the control mechanisms involved in these physical states.

Homeostasis

You probably think of the brain as the organ that produces intelligence, memory, and conscious thought, but it also controls aspects of metabolism like temperature control, respiration, heart rate, certain rhythms and cycles, and other maintenance and growth functions.

Sensory neurons detect the energy sources from both inside and outside your body. Receptors that detect energy include hair cells in the cochlea, receptors in the skin for pressure detection, and photoreceptors in the eye that detect light. There are also methods to detect internal body states that are interfaced in the CNS and the autonomic nervous system (ANS). Body temperature levels, blood pressure, and others are detected by these internal sensors that attempt to maintain these functions within an acceptable range of values. The acceptable range for these functions is typically referred to as *homeostasis*, a process that occurs largely without conscious awareness.

Fight-or-flight responses disrupt homeostasis to facilitate rapid energy use to deal with a threat. Typically these responses are the physiological components of stress. Chronic stress can be harmful, but no stress at all is also detrimental as the system is a like a muscle; it improves with exercise, but too much exertion can damage it.

From an evolutionary standpoint, homeostasis precedes the higher cognitive abilities that the brain is capable of. Even lower animals such as worms and insects have nervous systems that are able to regulate their internal states. A minor fluctuation or dysfunctions in these regulatory mechanisms can lead to health issues or worse.

In order for all the organs in your body to work efficiently, there must be some form of cooperation and regulation. For example, the volume of blood flow has to match the demand to move oxygen and nutrients where they need to go and to remove wastes; at the same time your respiration rate needs to match the demand for oxygen in your body. The ANS matches the activity levels of the different organs to balance these inputs and outputs. In addition, the ANS allows you to switch between various states of activity ranging from relaxation to activity, while at the same time maintaining a relative level of homeostasis over all of these states. The functions of the ANS are divided between the sympathetic nervous system (speeds things up in times of need) and the parasympathetic nervous system (slows things down).

ANS Contribution to Homeostasis

The sympathetic nervous system prepares one for the *fight-or-flight* response, and the parasympathetic nervous system slows the system down after the danger has passed. The sympathetic nervous system consists of neurons in the lumbar and thoracic portions of the spinal cord that use acetylcholine as their primary neurotransmitter. The sympathetic nervous system synapses on the sympathetic ganglia, a group of neurons running parallel to the spinal cord, and these neurons in turn synapse and target organs such as the heart. The sympathetic ganglia use norepinephrine (noradrenaline) as a primary neurotransmitter. The sympathetic nervous system also synapses on the adrenal medulla in the adrenal glands. The neurons in the adrenal medulla release epinephrine or norepinephrine into the bloodstream.

> Sympathetic nervous system effects include the inhibition of salivation, pupil dilation, constriction of blood vessels, increases in heart rate, relaxing the bladder, and others. Parasympathetic nervous system activation produces the opposite effects. Both of these ANS systems get sensory feedback from their targets via sensory receptors that detect functions such as blood pressure, temperature, etc. The feedback information is used to maintain homeostasis.

The parasympathetic nervous system has neurons located in the brain stem at the very posterior portion of the spinal cord, the sacral spinal cord. This system also uses the neurotransmitter acetylcholine. Parasympathetic nervous system neurons synapse on neurons near the target organs.

Hypothalamic Influence

The overall control of the ANS comes from higher levels, especially the hypothalamus (which means "below the thalamus"). The hypothalamus is a set of structures (nuclei) involved in controlling certain body states and functions, including hunger and thirst, circadian rhythms (fluctuations in body hormones that occur throughout the day), aspects of sexual activity, the release of hormones, and body temperature. **Figure 8-1** displays the hypothalamus and pituitary.

The hypothalamus can be stimulated from a wide range of areas in the brain including:

1. The orbitofrontal cortex, which is involved in a number of important functions, such as reactions to reinforcing or punishing stimuli, inhibition of activities, and emotional control.

2. The amygdala and hippocampus, associated with memories that in turn activate divisions of the autonomic nervous system by way of the hippocampus.

3. The medulla projections that carry information from the cardiac and digestive organs.

The hypothalamus has connections with numerous other areas of the body that are involved in the regulation of various body states. One of the ways the hypothalamus regulates these body states is through the secretion of hormones by way of the pituitary gland, which is at the base of the brain. The hypothalamus secretes *hypothalamic releasing hormones* that in turn stimulate the pituitary gland to release specific hormones into the bloodstream. Hypothalamic neurohormones include *vasopressin*, which is involved in kidney and cardiac functions; *oxytocin*, involved in reproductive and maternal behaviors in animals; and certain *growth hormones.*

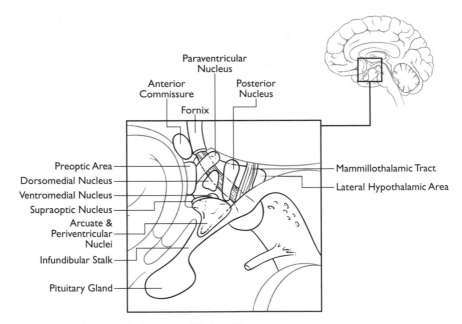

Figure 8-1: The Hypothalamus and Related Structures

Temperature Regulation

An example of how the hypothalamus regulates certain body states can be seen in a discussion of body temperature regulation. The hypothalamus detects fluctuations in body temperature by way of thermoreceptors on the skin, which detect the temperature of the external environment. These sensory receptors code for changes in temperature, and this information is relayed to the hypothalamus, which can in turn transmit impulses for corrective mechanisms to occur.

A critical area for temperature control is the ***preoptic area***, right next to the anterior hypothalamus. The preoptic area monitors body temperature by monitoring its own temperature. Cells in the preoptic area also receive inputs from temperature-sensitive receptors in the skin and spinal cord. Damage to the preoptic area impairs the ability to regulate temperature, meaning one can no longer shiver nor sweat effectively.

When you feel cold, information about this state may motivate you to put on a sweater or turn up the heat. If these actions are unsuccessful, your blood vessels constrict (vasoconstriction) and reduce blood flow to the skin, which minimizes heat loss (you also begin to look pale because of a lack of blood feeding the skin). If this doesn't work, you shiver, which is the result of your hypothalamus sending messages to activate the skeletal muscles to constrict. Shivering helps create heat and reduces heat loss. Another thing that might happen is that the hairs on your body could stand on end to trap the layer of air between the hair and skin, resulting in insulation of warmer air and the reduction of heat loss (the hairs are controlled by *arrector pili muscles* just under the skin).

For warmer temperatures the process moves in the opposite direction, and instead of shivering, your brain sends messages to the eccrine sweat glands under the skin to secrete sweat, which travels up the sweat duct through a sweat pore and onto the skin, resulting in evaporative cooling of the body. The signals from the hypothalamus can also result in the hairs on the skin lying flat, preventing heat from being trapped. Vasodilation increases blood flow through the arteries, redirecting blood into the capillaries in the skin. This helps to increase heat loss by convection and conduction.

Thirst and Your Brain

An earlier theory regarding thirst argued that it is the result of a dry mouth produced when the salivary glands in the mouth stop secreting fluid, thus cuing one to drink. However, empirical research discovered that if one were to put water in the mouth when one was thirsty, but prevented the water from reaching the stomach, one would still feel thirsty. Later theories resulted in identifying that thirst results from two major sources: osmometric thirst and volumetric thirst.

Osmometric Thirst

Eating salty foods makes a person thirsty. Within the cell there is a good deal of fluid, called intracellular fluid, that contains high levels of potassium, smaller levels of sodium, and smaller levels of other substances. Extracellular fluid, fluid outside the cells (but still in your body), has several sources. The first source, *interstitial fluid*, surrounds the cell bodies and is salty in nature; the second source, blood plasma (*intravascular fluid*), is found in the veins, arteries, and capillaries and suspends the cells and blood. Approximately 60 percent of your body weight is accounted for by these fluids: 40 percent of one's body weight is comprised of intracellular fluid and 20 percent of extracellular fluid.

The semipermeable membranes surrounding the cells do not easily allow the passage of sodium, and so water is drawn from the intracellular fluid to the extracellular fluid by osmosis. Osmosis occurs when water moves through a semipermeable membrane from an area that has a low concentration of solutes to an area where there is a high concentration of solutes (a solute is simply a substance dissolved in another substance, so the fluid inside the cell flows out because the fluid outside the cell has higher levels of solutes). Eating salty foods results in more sodium in the extracellular fluid than in the intracellular fluid, which produces osmotic pressure, so the flow of water moves from inside the cells to the fluid outside the cells. The movement of water leads to a reduction in the concentration of sodium across a membrane and shrinks the cell's body (dehydration). The brain is equipped with osmoreceptors that inform the CNS that this dehydration is occurring. These receptors are believed to be in the lateral preoptic area. When you eat, this leads to water being diverted from the rest of the body into the stomach and small intestine, to be used in digestion. Once the food is absorbed, it increases the solute concentration of the blood plasma and thus induces osmometric thirst.

The signal to stop drinking comes from messages sent from the stomach to the brain that record how much the stomach and small intestine have distended. If the message somehow came from the cells that had hydrated, you would drink far more than you need to.

Volumetric Thirst

Dehydration that occurs outside the cells as opposed to inside the cells (osmometric thirst) is known as volumetric thirst, because this thirst is induced by a reduction of blood plasma (some texts refer to this as hypovolemic thirst). Obviously a large loss of blood would lead to a great loss of extracellular fluid and lead to an increase in thirst.

Volumetric thirst can also result from low levels of salt in the diet, which produces a loss of extracellular fluid by a movement of water from the extracellular fluid into the cell bodies by osmosis. An important structure in detecting this loss of fluid is the kidney. If blood flow to the kidney is reduced, the kidney releases an enzyme called *renin*. Renin produces two hormones: angiotensin I and angiotensin II. These hormones appear to be important in your awareness of volumetric thirst.

Excessive perspiration could lead to volumetric thirst as the body loses (primarily) extracellular water and results in an imbalance in the extracellular fluid.

The walls of blood vessels also have receptors called baroreceptors, which detect blood pressure changes. A loss of extracellular fluid can produce a drop in blood pressure that would be detected by these baroreceptors and initiate volumetric thirst by way of your brain. Baroreceptors are essential for homeostasis and are pressure-sensitive nerve endings in the walls of the atria of the heart, the aortic arch, and the carotid sinuses. These stimulate central reflex mechanisms allowing for adaptation to changes in blood pressure by way of changes in heart rate or by vasodilation or vasoconstriction.

Hunger

There is a popular theory in textbooks called the set point theory. Set point theory in eating and weight control is based on the premise that every person has an inherent system that controls precisely how much fat they should carry or how much they should weigh. Some are purported to have a higher set point for these features than others, so weight management would vary by the individual.

There are obvious weaknesses to the set point theory of hunger and eating-related behavior. First, epidemics of obesity in industrialized nations are counterintuitive to the idea that appetite is controlled by a set point. If a set point exists to balance food intake and energy expenditure, epidemics of obesity should not exist.

Secondly, set point theories are not consistent with the notion that in the wild animals will often eat large amounts of food and accumulate higher levels of body fat during the fall to prepare for a lack of available food during the winter, or when in captivity, where they have no need to save up energy stores. If eating and hunger are controlled by a set point that balances food intake and energy output, this behavior should not occur.

Finally, there is a large body of research that indicates that cues like television, smell, taste, social learning, stress, etc., increase the feeling of hunger in people. Again, if hunger is controlled by a set point that balances energy output with food intake, these environmental cues should not have a significant impact on feeding.

> A current theory regarding how the brain works is the notion that the adult human brain is quite capable of adapting to changes in the environment (*plasticity*), compared to older notions that the structure of the adult human brain does not change much in response to the environment. Set points would not be very plastic. Nonetheless, the set point assumption regarding hunger and eating still remains a dominant theory in many texts.

The Hypothalamus and Eating

Much of the earlier research with rats indicated that the hypothalamus controlled aspects of feeding and satiety (you can still find mention of this research in many texts). Feeding behaviors were believed to be controlled by the lateral hypothalamus (LH), and the sensation of satiety was believed to be controlled by the ventromedial hypothalamus (VH). More recent research has indicated that the primary role of the hypothalamus is regulating energy metabolism and not regulating eating behavior directly. The earlier research with rats that showed that lesions of their VH led to obesity was misinterpreted. The earlier research concluded that the rats became obese because they overate. However, research now indicates that lesions in the VH led to increases in blood insulin levels in rats, which increased stored body fat in them and decreased the breakdown of fat that normally would be utilized as an energy source. Thus, the food that the rats with bilateral VH lesions ate was converted to fat quickly, so the rats had to keep eating to have enough calories in their systems to meet their energy requirements (stored fat in the rats' systems could not be broken down). So what really happened to the rats with the VH lesions was that they overate because they became obese!

Subsequent evidence on the effects of LH lesions has also indicated that the LH may not have a direct involvement in feeding behavior. Earlier research indicated that lesions of this area produced total stoppages of eating and drinking behaviors in rats. Researchers believed that these lesions were evidence that the LH was involved in feeding behaviors. Subsequent research has indicated that damaging the LH results in a lack of responsiveness to sensory inputs and motor disturbances in the animal, of which eating and drinking are only two examples. Therefore, the theory that the LH was specifically involved in feeding and drinking is not valid. So it appears that damage to the hypothalamus *indirectly* affects eating behaviors.

> The positive-incentive theory states that eating is driven by the anticipated **pleasure** of eating a particular food. If this theory were valid, it should also explain other similar behaviors that have similar neuro-representations, such as drug addiction. As it turns out, this has been a major theory used to explain addiction.

A number of *positive-incentive* theories have been developed as a result of recent research to explain why animals eat (or at least why warm-blooded animals eat). These theories generally propose that people are not drawn to eat because of an internal energy deficit, but instead are drawn to eat by the anticipated pleasure that eating gives them (termed the *positive-incentive value*). These theories assume that there is an evolutionary drive to eat because people crave the pleasure of eating. Warm-blooded animals have been shaped by pressures of unexpected food shortages to take advantage of food when it is available and eat it because they need a continuous energy supply to maintain their body temperatures. These theories are more consistent with external cues in the environment that trigger eating and learned preferences and aversions for different types of food. Thus, eating behaviors are probably influenced by learning, the reward mechanisms in the brain, and physiological feedback from the gastrointestinal tract.

Feeling Hungry or Full

The evidence has mounted that the gastrointestinal tract releases *peptides,* short chains of amino acids that function as either neurotransmitters or hormones. When you eat food, the gastrointestinal tract interacts with receptors that release peptides into the bloodstream. Several of these peptides have been shown to bind to receptors in the brain and reduce food intake (interestingly many of these peptides bind in hypothalamic areas involved in energy metabolism). There are also several peptides, called *hunger peptides,* which increase appetite. These appear to be synthesized in the brain as well (again, many of these are synthesized in the hypothalamus). Since there is a large number of these hunger and satiety peptides, it is obvious that feeding behaviors react to multiple signals. The discovery of these peptides has led to a renewed interest in the role of the hypothalamus in hunger. However, the hypothalamus is most likely a part of a larger circuit, not yet well defined, that is involved in feeding behaviors. Another substance that also acts as a neurotransmitter—but outside the CNS—and is important in feeding behaviors is serotonin. Serotonin appears to increase feelings of "fullness" or satiety.

Eating Disorders

Two major eating disorders are recognized: *anorexia nervosa* and *bulimia nervosa.*

Anorexia is a disorder of consumption such that the person (most often female) experiences life-threatening weight loss and other health issues. People with anorexia have a distorted view of their body and see themselves as fat even though they may be severely underweight. This leads to severely restricted eating behaviors despite an obsession with food-related activities such as cooking. They have a high rate of suicide and nearly 10 percent die of starvation.

Bulimia is characterized by periods of binging (consuming large amounts of food in a very short period of time), followed by periods of voluntary purging, which can occur by self-induced vomiting, use of laxatives, or extreme exercise. People with bulimia are typically not underweight.

There are a good number of theories as to why these disorders occur, such as alterations in brain systems including the hypothalamic-pituitary-thyroid mechanisms,

alterations in catecholamine metabolism (the neurotransmitters epinephrine, norepinephrine, and dopamine), and activities and alterations in endogenous opiate activity
(neurotransmitters that control pain). Specific causes for eating disorders have not been
identified. Treatment for these disorders is most effective if initiated early in the course
of the disorder; however, relapse rates are high. Some psychiatric medications appear to
enhance recovery in individuals with these eating disorders, but medications are most
effective if combined with psychotherapy.

Obesity

At the time of this writing, obesity is not recognized as a formal eating disorder, but then
again why should it be? Obesity refers to one's body size and composition, not to eating
habits. The relationship between obesity and health is well documented. From an evolutionary perspective, obesity is often understood as a mechanism that was designed to deal
with food shortages. However, in America it is typically believed that a person should eat
three regular meals a day; food should be the focus of many social events; fatty substances
like butter or sweet and salty substances should be added to food to make them more fla-
vorful; and meals are best served in courses. People who live in industrialized nations live
in an environment with an endless variety of food that has the highest positive-incentive
value, which coincidently includes a high calorie composition.

> Overweight and obesity are defined as excessive fat accumulations resulting in
> health risks. A crude measure, the body mass index (BMI), is a person's weight
> (in kilograms) divided by the square of his height (in meters). A BMI of 30 or
> more is considered obese, whereas a BMI between 25 and 30 defines overweight.
> However, this index cannot consider percentages of body fat.

Several factors contribute to the rising rate of obesity in industrialized countries,
including genetic factors (over 100 different loci on human chromosomes have empirical links to obesity), differences in food consumption, the makeup of foods consumed in
these countries, and differences in activity levels. A promising discovery in understanding and treating obesity was the discovery of *leptin*, a peptide hormone that is released
by fat cells. Leptin acts on receptors in the hypothalamus to counteract a peptide that
stimulates feeding (neuropeptide Y, secreted by cells in the hypothalamus and gastrointestinal tract). It also counteracts other neurotransmitters that stimulate feeding, such

as anandamide, and promotes the synthesis of the appetite suppressant α-MSH (alpha-Melanocyte-stimulating hormone). Leptin is a circulating signal that reduces appetite, and obese individuals typically have high concentrations of leptin in their blood. This has led researchers to hypothesize that perhaps people who suffer from obesity are somehow resistant to the effects of leptin. If this is the case, then it could be possible to develop medications that would combat this resistance. This research is ongoing.

A 200-pound man running at a seven-minute-mile pace burns 162 calories/mile. To burn a pound of fat, he would have to run 21.6 miles. This is why exercise alone is not an effective way to lose weight. If the same person cut 500 calories a day from his diet, he would achieve the same effect as running three miles daily for a week.

Other treatments for obesity include diet and exercise programs, gastric surgical procedures, such as a gastric bypass surgery or adjustable gastric band procedures, medications, and a number of other remedies found in the popular media. There has been some research that indicates that the use of drugs that enhance the effects of the neurotransmitter serotonin can reduce appetite in some obese individuals. However, the medications used to enhance serotonin in early studies were also associated with some severe side effects, such as heart disease. This research is also ongoing. At this time, the best cure for obesity appears to be sound dietary habits.

SEX

THOUGHTS ABOUT SEX NEED THE BRAIN. Engaging in sexual activity also needs the brain. Thinking about sex and doing it at the same time requires a brain (believe it or not!). Intentions, desires, and actions result from an interaction between biochemistry and neuroanatomy. But still, there are so many unanswered questions about sex. Why do some people prefer their own gender as sex partners? Why do some people become sex offenders? How is the brain different in males and females? Take a cold shower and read on.

Sex Hormones

Hormones influence sexual activity by affecting the development of the physiological and behavioral characteristics of males and females throughout their lifespan *and* by activating behaviors involving reproduction in sexually mature individuals.

The organs in the body whose primary function is the release of hormones are referred to as *endocrine glands*. However, like many designations in human anatomy, this definition is not without its exceptions. Other organs that do not function to *primarily* release hormones in the system, such as the liver and stomach, also release hormones in the body and are also technically part of the endocrine system.

Glands come in two types: *exocrine glands*, which release chemicals into ducts that then carry the chemicals to the surface of the body (e.g., sweat); and *endocrine glands*, which release hormones directly into the circulatory system. Typically, when people speak of the endocrine system, they are including the testes and ovaries, pancreas, adrenal glands, thymus glands, parathyroid gland, thyroid gland, pituitary gland, hypothalamus, and the pineal gland.

> The **organizing effects** of sex hormones occur mostly at sensitive stages of development. In humans these occur before birth and determine the biological gender of the fetus. The **activating effects** of sex hormones occur at any time in life when a hormone activates a specific response. The effects of activating hormones may last for hours or for months, but they do not last indefinitely.

Hormones

Hormones are chemicals released into the blood stream by glands, whereas neurotransmitters are confined to the CNS. Some substances act as both. Hormones can be designated by three types:

1. *Amino acid derivative hormones*, synthesized from an amino acid molecule.

2. *Peptide and protein hormones* are chains of amino acids. Peptide hormones are short chains of amino acids, whereas protein hormones are longer chains of amino acids.

3. *Steroidal hormones* are all synthesized from the fat molecule cholesterol.

The hormones involved in sexual development and sexual behavior are steroidal hormones. Steroidal hormones can bind to the cell membrane or penetrate the cell membrane and bind to receptors in the cytoplasm or nucleus of the cell. By penetrating the cell membrane, steroidal hormones directly influence gene expression in the cell. Steroidal hormones are produced in the gonads (testes or ovaries, which also produce sperm and ova, respectively). The two main classes of hormones released by the gonads are *androgens* (testosterone is the most common) and *estrogens* (estradiol is the most common). A third class of steroidal hormones, progestins, is also released (the most well-known of these is progesterone, which in women during pregnancy, stimulates the uterus and breasts). Ovaries and testes release both androgens and estrogens; however, ovaries will tend to release more estrogens and testes more androgens. The outer layer of the adrenal glands (called the adrenal cortex) also releases small amounts of steroidal hormones into the system.

> It is a myth is that androgens are "male hormones" and estrogens are "female hormones." Androgens do not produce more "maleness" in a person and vice versa. Both types of hormones are found in either gender. Typically, the relative proportion of each differs in the genders.

Males and females have the same hormones; however, the hormones are not present at the same levels in males and females and may not perform the same functions. One of the main differences is that the levels of hormones in females are subject to a cycle that repeats itself about every twenty-eight days (the menstrual cycle), whereas the levels of hormones in males tend to remain relatively stable.

The master gland in the body is the pituitary gland (pictured in **Figure 8-1**). The pituitary gland primarily releases *tropic hormones*, hormones that primarily influence the release of other hormones. The pituitary gland is typically divided into the anterior pituitary and the posterior pituitary. Technically, the anterior pituitary functions to release tropic hormones, and the posterior pituitary releases hormones such as oxytocin (involved in lactation and uterine contractions) and vasopressin (which stimulates water retention, contracts arterioles, and may influence male aggression). Therefore, many neuroanatomists designate the anterior pituitary gland as the "master gland."

Both the anterior and posterior pituitary glands are under the control of the hypothalamus. Vasopressin and oxytocin are synthesized in the hypothalamus and transported to the posterior pituitary where they are stored until they are needed. The hypothalamus does not have any neuronal connections to the anterior pituitary gland. Instead, the hypothalamus affects control over the anterior pituitary by way of hormone release (by way of *releasing hormones* and *release-inhibiting hormones* that stimulate or inhibit, respectively, the release of hormones from the anterior pituitary).

> Of course males and females differ physically and in their sexual behaviors. They also differ in characteristics that may be indirectly related to reproduction. Males of most mammalian species tend to be larger and more aggressive than females, whereas females live longer and devote more attention to the care of the young. Humans are among the few mammals where males help care for their young.

The release of hormones typically occurs in short periods several times a day. These releases typically consist of gushes or surges. This means that there often are fluctuations in levels of circulatory hormones, and the release of hormones is regulated by three major types of signals. With the exception of the anterior pituitary gland (regulated by the hypothalamus), all endocrine glands are regulated by signals from the nervous system. The pituitary and pineal glands are regulated by cerebral neurons, whereas in the peripheral nervous system, glands are regulated by the autonomic nervous system. The environment plays an important part in the regulation of many hormones; for example, breeding seasons for many animals are regulated by the weather.

Hormone release can also act as a signal to trigger the release of other hormones. As previously discussed, the hypothalamus releases hormones that in turn influence hormone release from the anterior pituitary. In addition, the presence of other chemicals, such as sodium and glucose levels in the blood, can influence the release of certain hormones. Most people know that insulin is released from the pancreas in response to high blood glucose levels. The release of insulin leads to a reduction in blood glucose.

Hormones and Reproductive Behavior in Females and Males

In female mammals (other than humans) surges of estrogens result in *estrus*, a time period during which the female is receptive to sexual approaches by males of the same species. During this period, the female may produce pheromones that increase her sexual attractiveness to males and signal that she is fertile. There is a relation between the cycle of estrus and hormone release in females that begins with a gradual secretion of estrogen by the follicles (the ovarian structure where eggs reach maturity), which is followed by a surge in progesterone as the egg is released.

In some primates, such as humans, cycles of increased fertility appear to have little relation to the motivation to mate. There have been studies that have tried to determine the role of estradiol and sexual interest in human females; however, these studies often have conflicting results, with some indicating that there is an increase in sexual interest, while other studies indicate little or no change in sexual interest during hormone release.

Studies investigating the role of androgens and female sexual interest in humans have demonstrated some positive effects. For example, in studies of women, their responses on measures of sexual motivation appear to be related to their testosterone levels. Moreover, in studies of women following menopause, injections of testosterone have appeared to increase their sexual interest. However, more research needs to be done.

Males

Orchiectomy is the surgical removal of one or both testes. Early studies of sex offenders who had agreed to be castrated indicated that there was a marked and almost complete lack of recidivism. In one large study, nearly half of the men lost all interest in sex within a few weeks of the operation. Some were unable to achieve an erection but still had some sexual interests, and a few were able to continue to copulate successfully. Other studies and clinical observations appear to support the notion that testosterone is important in male sexual interest and in male sexual behavior (such as studies looking at how injections of testosterone affect sexual behavior in older men). However, testosterone levels appear to be unrelated to sexual interest in healthy men with normal sex drives, as testosterone injections do not increase their sex drive. Of course, testosterone replacement will not treat sterility in men.

Chemical castration involves the administration of antiandrogen drugs or certain antipsychotics every three months to sex offenders as a form of probation. Studies have indicated similar effects to actual castration; however, groups like the ACLU believe administration of these drugs is a form of cruel and unusual punishment. Several states have experimented with chemical castration as an alternative to lengthy prison sentences for these offenders.

Development and Puberty

One attitude that many people seem to have is that men and women are fundamentally more different than they are alike. Obviously males and females are different (thank goodness). However, as you should have surmised by now, the idea that males have male hormones that make them do "male" things and females have female hormones that make them do "female" things is not accurate. With respect to brain differences, the brains of men are about 15 percent larger than those of women (but remember, brain size is not necessarily related to increased intelligence).

During development in the womb, hormonal release results in the development of some masculine or feminine physical characteristics in embryos. Recall that sex hormones are steroidal hormones and that all steroidal hormones are derived from cholesterol. This means that all sex hormones are similar in structure. Under the influence of the enzyme *aromatase*, testosterone converts to the hormone estradiol (this process is called aromatization). In the process of embryonic development, some human sexual characteristics are determined by the hormone estradiol. The aromatization theory purports that the default program in developing embryos is female, but that the influence of hormones differentiates male and female embryos. The aromatization hypothesis states that embryonic brains become masculinized by estradiol that has been aromatized by testosterone, and that in female embryos, a substance called *alpha fetoprotein* deactivates this process. However, there are a number of other influences that determine the gender-related physical characteristics of the developing human, such as the sex chromosomes (which appear to influence brain development independent of hormonal influences), various differences between the brains of males and females that develop at different stages, and sexual differences in the brain that appear to be due to different mechanisms in different species of mammals. (Aromatization may be more important in rodents than in primates.) In summary, the differences in the brains of males and females emerge at different stages of development and under different types of genetic and hormonal influences.

Puberty

Children tend to have low levels of circulating gonadal hormones and just a few differences in general appearance until puberty. At puberty, which is the transition between childhood and adulthood where the person becomes sexually fertile, certain secondary sex characteristics develop. These include the development of the sex organs, breasts in females, facial and body hair in most males, and the other typical characteristics associated with males and females. Puberty is associated with hormone release by the anterior pituitary gland. Increases in the release of gonadotropic and adrenocorticotropic hormones lead to the gonads and adrenal cortex releasing their hormones, which in turn initiates the maturation of the genitals and the differentiation of secondary sexual characteristics. In general, during puberty androgen levels are higher than estrogen levels in males, whereas estrogen levels are higher than androgen levels in females. Currently, the average age that puberty begins for males in the United States is about eleven and a half years of age and at around ten and a half years of age for females. However, about a century and a half ago, these averages were sixteen and fifteen years old. Most attribute this to changes in diet, medical treatments, and even social conditions (another illustration of *plasticity*).

Perinatal hormones also influence the development of behavior. In earlier experiments on animals (mostly rodents), injections of testosterone could lead to mounting behavior in females, and a lack of early exposure to testosterone in male rats feminized their mating behavior. Timing was critical in these experiments, and the effects of testosterone injections to "masculinize" the female rats appeared to be restricted to the first ten or eleven days following their birth. Less is known about the effects of hormones on other behaviors and how they relate to males and females. Certainly in humans there is a combination of biological and social effects that determine the behavior of the person's gender. Moreover, as strictly masculine and strictly feminine behaviors become less well-defined in society, more and more people no longer conform to stereotypical patterns of gender-appropriate behavior. The bottom line is that much research is still needed regarding the effect of hormones on "masculine" or "feminine" types of behaviors.

A *gender role* contains a set of social and behavioral expectations considered appropriate for either a male or a female in social circles or in interpersonal relationships. Gender roles exhibit a wide variation between cultures and even vary within the same culture over time and within different contexts.

Sexual Dimorphism in the Human Brain

Sexual dimorphism refers to differences between males and females of a given species. Nuclei in the suprachiasmatic, preoptic, and anterior hypothalamic regions differ in men and women. One discovery that appears to be consistent across the mammalian species is a nucleus (a collection of neurons) in the medial preoptic area that is larger in males than in females. This was originally discovered in rats in the 1970s, and oddly enough, this cluster of cells was called "the sexually dimorphic nucleus." In rats, the structure is the same size for males and females at birth, but its rate of growth in males is much faster. The rate of growth of the sexually dimorphic nucleus appears to be mediated by estradiol that is aromatized by testosterone. The size of the male rat's sexually dimorphic nucleus appears to be related to testosterone levels and the rat's sexual activity; however, the specific function of the structure is unclear. It appears that the medial preoptic area of the hypothalamus (which includes the sexual dimorphic nucleus) is involved in sexual behavior in males, as when this area is destroyed, sexual behavior ceases in male rats, but not in females. Likewise, electrical stimulation of this area initiates mounting behaviors in animals.

Determinants of Gender Identity and Sexual Orientation

Most boys and girls living in Western countries start having their first feelings of sexual attraction at about age ten, which is also about the age that puberty sets in. This may also be related to the maturation of the adrenal cortex, which also matures at about age ten. There is more and more research that is implicating one's genetic makeup as a major contributor to sexual orientation. The *concordance rate* for a particular behavior relates to the probability that one person will have a given trait when another person has that trait. For instance, identical twins share the exact same genetic makeup, whereas fraternal twins share about half of their genetic makeup (the same as any other siblings). In order to determine how much a particular trait is due to genetic influences, comparisons of identical and fraternal twins are often made. The concordance rates for sexual orientation range from 20–50 percent in identical twins and between 9–22 percent in fraternal twins, depending on the study.

What this statistic means, or is at least taken to mean, is that genetic factors account for up to about half of the expression of sexual orientation. Unfortunately this interpretation is wrong; however, it does suggest that there is a large contribution of genetic factors to sexual orientation. Additionally, it suggests that there is also a large contribution of other factors that determine one's sexual orientation, such as the environment, because the concordance rates in these studies are never 100 percent for identical twins. One of the things that should be kept in mind about studies that tie genetics and behavior together is that everything has some genetic contribution to it; however, genetic influences require interactions with the environment in order to be expressed. So the "nature versus nurture" debate that often comes up in these discussions is really not relevant.

> *Gender identity* refers to what particular gender a person defines oneself as: either male or female. *Sexual orientation* refers to what gender a person is sexually attracted to or would want to have sexual relations with. A person can believe they are female and want to have sex with females or with males, and vice versa. The two concepts are not necessarily designed to be dependent on one another.

There are also many studies that attempt to identify differences in certain brain structures of heterosexual and homosexual participants; however, there have been no reliable differences found in the brains of these two groups at the time of this writing. Likewise, heterosexuals and homosexuals do not appear to have differences in levels of circulating hormones. The notion that sexual preference is fully due to choice also appears not to be reliable, because in many cases sexual preferences are seen very early in children. Studies using rodents have indicated that perinatal castration in males increases their preference for males as sexual partners when they are adults, and prenatal testosterone exposure in females increases their preference for female sex partners when they are adults. The problem with these studies of course is that "sexuality" does not transfer well between rodents and humans, and of course studies like this would be unethical to do on humans (you could not castrate or expose human embryos to hormones in an experiment). There is sparse evidence that perinatal hormones contribute to sexual orientation in humans.

> There is also a theory known as the *fraternal birth order effect* that observes that the probability of a male being homosexual increases with the number of older brothers that he has. This has been explained by the hypothesis that in some mothers—no one knows which mothers—there is a progressive immunity that has developed to masculinizing hormones in male fetuses, and as more children are born, there is a greater chance that males may be homosexual.

Gender Identity

Most often one's gender identity is the same as one's anatomical sex, but not always. *Gender identity disorder* is a condition in which the individual believes that his or her sexual identity is opposite his or her anatomical or genetic presentation. The causes of gender identity disorder are not known, although it is assumed that this disorder is related to genetic and hormonal influences. There are treatments for gender identity disorder; typically sexual reassignment surgery is offered. This requires an intense psychiatric and psychological evaluation of the person, their living as the opposite sex before surgery for a specified period of time (often a year), and then surgery and hormone therapy.

Thinking about Sex

At the anatomical level of the brain, the areas that appear to be activated when thinking about and executing sexual behaviors are the preoptic area of the hypothalamus and the amygdala. These regions are part of the *limbic system*, a system involved in emotional behaviors and memory. Each of these brain regions contains receptors that are sensitive to sex hormones. In males who have suffered strokes that affect these areas, there is a reduction in libido and sexual performance in some studies, but in other studies such patients become sexually hyperactive. The center of sexual initiation in females may involve the hypothalamus and the thalamus.

The amygdala is important in emotions, and so a role for it in sexual behaviors should not be surprising. Human patients who have had their temporal lobes removed, and thus their amygdalae removed, have shown evidence of excessive sexual behaviors. Such findings may signal that the amygdala is somehow responsible in the inhibition of certain behaviors through its connections with the orbitofrontal cortex. Studies of orbitofrontal cortex damage in individuals also indicate a loss of inhibitions, often a loss of inhibitions of a sexual nature. Typically, the right hemisphere has showed the most consistent activation during sexual arousal in the aforementioned brain regions as well as in other regions of the brain.

Sex Offenders

Rape is not a sexually motivated crime, but instead is a crime of aggression, so it will not be dealt with in this book. One of the most researched and still misunderstood behaviors is that of *pedophilia*, sex crimes where adults abuse children. Pedophiles have been rarely studied extensively in postmortem studies, but there have been some differences observed in the brains of pedophiles and normal controls upon autopsy. For example, one study compared the brains of fifteen convicted pedophiles to fifteen normal control brains postmortem. The most significant difference between the brains was in the area of the right amygdala, which was significantly smaller in pedophiles than in normal controls. There were several other differences in brain structures noted in the study, suggesting perhaps that pedophiles suffer from compulsions related to pathological or developmental insults of the brain. However, these studies typically use averaging techniques where the changes in the structure of the brains between groups are averaged over each group. Moreover, a study with only thirty subjects cannot be generalized to the population. Nonetheless, the study does propose some interesting questions about pedophiles and their actions. Are pedophiles the victims of a brain disease/disorder in which they are compelled to act, or are they engaging in volitional action?

THE SLEEPING BRAIN

No MATTER HOW MUCH a person loves eating or sex, one could never spend as much time doing them as sleeping. An average person may sleep 200,000 hours in her lifetime. What is the function of sleep? If sleep is a form of cessation of activity like catching your breath, there is no reason to search for the neural mechanisms responsible for sleep. However, if sleep is a special state of consciousness serving a particular function, then researchers should seek answers as to how it is regulated.

Bodily Rhythms

The environment in which you live is full of routines and cycles. Of course the most common cycle is the day-night cycle that occurs over every twenty-four hours. Most of the animals that are exposed to this light-dark cycle have adapted to this timed routine such that they have fairly regular schedules of sleeping, waking, eating, etc. *Circadian rhythms* are these daily cyclical adaptations, which can be biological or functional in nature.

The most obvious circadian rhythm is the sleep-wake cycle that people and animals adhere to; however, many other biological processes, such as body temperature and hormonal levels, adjust themselves to a circadian rhythm. Moreover, circadian rhythms such as the sleep-wake cycle still endure in conditions where there is no light-dark cue. In laboratory conditions when animals or people live in total dark or total light, it appears that the cycle adjusts to regularities in the animal's environment such as regular meal times, regular social interactions, and other events that are predictable.

> When a person works on the night shift, he tends to sleep during the day. Daytime sleep periods are about an hour to two hours shorter than nocturnal sleep periods. Stage II sleep and REM sleep are typically more affected than other stages, and this leads to problems with alertness and fatigue. It is not uncommon for shift workers to become obsessed with sleep.

It also appears that the circadian rhythm can deviate from the twenty-four-hour cycle. Animals exposed to eleven-and-a-half-hour cycles of light and dark will display a circadian rhythm that adjusts to a twenty-three-hour cycle. When environments are devoid of events that can serve as cues for circadian rhythms, it appears that these rhythms are still maintained. These *free-running rhythms*, as they are termed, have *free-running periods* that vary in length and are only slightly longer than twenty-four hours, suggesting that people have an internal biological clock that approximates the twenty-four-hour cycle.

Neural Mechanisms of Circadian Rhythms

An internal *circadian clock* is the biological timing mechanism that is responsible for the circadian rhythms in people and animals. The area of the brain that has been linked to circadian cycles is called the *suprachiasmatic nucleus*. The suprachiasmatic nucleus is located in the medial area of the hypothalamus on both sides of the brain, opposite the *optic chiasma,* which is the point where the optic nerves cross over to the opposite hemisphere of the brain. The suprachiasmatic nuclei appear to be involved in the setting of light-dark cycles of circadian rhythms, by way of neural projections from the optic nerves. It appears that axons project from the optic nerve into the area of the optic chiasma and project to the suprachiasmatic nuclei. These projections have been termed *retinohypothalamic tracts*; therefore, visual information is the cuing mechanism for this route, and the suprachiasmatic nucleus appears only to be involved in circadian rhythms that are cued by light-dark cycles, not those triggered by other stimuli. When animals in experimental conditions have this area removed or lesioned, other environmental cues, such as food, can still trigger circadian rhythms, but light-dark cycles do not. So while the suprachiasmatic nuclei do appear to be involved in the light-dark circadian rhythms, there obviously are other areas of the brain that are also involved in regulating these cycles.

> Biorhythms are attempts to track various aspects of functioning—emotional, intellectual, and physical cycles—using mathematical techniques. Empirical research has indicated that these techniques and their assumptions are not reliable, and most scientists find no support for them. Chronobiology is the study of biological cycles, such as the circadian rhythm.

One mechanism that may contribute to circadian rhythms is genetics. Certain genes are involved in circadian rhythms, and many of the same gene types have been identified in many different species of animals. More recent research has indicated that molecular circadian timing mechanisms, similar to the type in the suprachiasmatic nucleus, exist in many cells of the body, and that these are affected by certain hormones and neural influences from the suprachiasmatic nucleus.

Stages of Sleep

The amount of sleep that people get is variable from one person to the next, and the common prescription is everyone should get eight hours a night. Most people who live to be seventy-five years old will have spent almost twenty-five years sleeping. There are many changes that occur during sleep, and most of the studies that investigate the processes occurring during sleep use either EEGs (that measure brainwave activity), electro-oculograms (EOG, that measure retinal activity), or electromyograms (EMG, that measure electrical impulses of muscles) to record their findings. EEG findings have determined that there are four separate stages of sleep that are appropriately named stage I, stage II, stage III, and stage IV. There is also another stage, REM sleep.

Before you fall asleep, there is a period of relaxed wakefulness that produces EEG waves known as *alpha waves* that have a frequency of about 8 to 12 waves per second. Alpha waves are characteristic of a relaxed state. Once one enters stage I sleep, one produces EEG waves that are irregular, jagged, and low-voltage, indicating that brain activity is declining. These waves are similar to those produced during alert wakefulness but are much slower.

The most prominent EEG characteristics of stage II sleep are sleep spindles and K-complexes. A sleep spindle is a burst of 12 to 14 Hz waves that last at least a half second (typically these last for two or three seconds). The K-complex consists of a sharp, high amplitude, negative wave followed by a smaller, slower, positive wave. K-complexes are most common during stage II sleep, but sudden stimuli can evoke these in other stages of sleep.

Stage III sleep reveals the presence of *delta waves,* which are the largest and slowest brain waves recorded on an EEG. Delta waves are also predominant in stage IV sleep, thus leading to stages III and IV sometimes being called *slow-wave sleep.*

As you fall asleep and move through the stages, your breathing rate, heart rate, and brain activity become slower than they were in the previous stage, and the percentage of slow, large amplitude waves increases. These slow brain waves indicate that neural activity is becoming highly synchronized, whereas during wakeful states, neural activity is a less synchronized response to a number of external and internal stimuli. Once at stage IV sleep, you remain there for a while and then cycle back through stages III and II, but instead of going to stage I, you enter a new stage called *REM sleep* (rapid eye movement sleep). Some researchers prefer to distinguish only between REM sleep and non-REM sleep (all of the other stages).

REM sleep is obviously associated with rapid eye movements, but also with a loss of muscle tone in the body's core and high-frequency, but low amplitude, EEG waves. Cerebral activity increases such that REM sleep is similar to waking states. Autonomic nervous system activity also waxes and wanes during REM sleep. There can also be occasional twitches in the extremities during REM sleep and in many males there is penile erection. REM is associated with dreaming, and for many years some texts suggested that dreaming only occurs during REM sleep; however, even early studies of dreaming and REM sleep indicated that people could remember dreams from non-REM sleep, even if this was infrequently. So while it appears that a great deal of dreaming does occur during REM sleep, it is also certain that some dreaming occurs in other stages.

Brain Areas Involved in Sleep

The hypothalamus is one of the main areas of the brain that appears to be involved in sleep. Very early discoveries based on clinical data indicated that people who had damage to the posterior hypothalamus and adjacent areas experienced problems with excessive sleep. Other studies of individuals who had damage to the anterior hypothalamus experienced difficulty sleeping; thus, early findings based on clinical studies of patients suggested that the anterior hypothalamus is involved in promoting sleep and the posterior hypothalamus is involved in wakefulness.

The sleep-wake cycle appears to be a function of an area of the brain called the *reticular activating system* or *reticular formation*. The reticular formation is an extensive network of nuclei and nerve pathways located throughout the brain stem. This series of structures connects both motor nerves and sensory nerves to and from the spinal cord and the cerebellum, and the reticular formation also has widespread connections to the cerebrum. The neurons in the reticular formation also have extensive connections with other areas of the brain and perform numerous other functions. It has been estimated that a single neuron in the reticular formation may have connections with as many as 25,000 other neurons. This important center of the brain is involved in alertness, sleeping, attention (for example, filtering out stimuli that are unimportant and selecting stimuli that are important for further scrutiny), pain control, cardiac functions, and other functions.

> The reticular formation is involved in alertness and the ability to filter out irrelevant stimuli. The reticular formation also produces acetylcholine, an important neurotransmitter. Bilateral damage to this area may result in coma or death. Neuroimaging studies have shown abnormal functioning in the reticular formation in people with chronic fatigue syndrome, suggesting that this area has some interplay with the fatigue associated with certain disorders.

The posterior portion of the reticular formation appears to be involved in the production of REM sleep. There are several sites at the posterior portion of the reticular formation that each appear to be involved in a separate function relating to the characteristics of REM sleep. One site of the posterior reticular formation appears to be involved in rapid eye movements, another site in the reduction of muscle tone, and so forth.

The Dreaming Brain

Dreams have intrigued people for centuries. There are numerous stories of dreams having the ability to predict the future or to have some deep hidden meaning. Freud believed that dreams were expressions of a person's unconscious needs and drives, and he used dream interpretation as a part of his psychoanalysis. Nonetheless, there is no evidence that dreams are premonitions or that the Freudian theory of dreams is valid. More current theories, such as the *activation-synthesis theory*, suggest that dreams result from a mass of information supplied to the cortex during REM sleep and that this information is random. The dream is an attempt by the brain (cortex) to make sense of these signals.

Some people claim that they never dream; however, in clinical studies these people have as much REM sleep as other people who do recall dreams, and if they are awakened during REM sleep, they do report the aspects of dreams. Another myth about dreams is that dreams only last a few seconds or less, but again clinical research has refuted this belief. The research suggests that dreams run on real-time, so if your dream feels like it lasted for thirty seconds, it probably lasted pretty near thirty seconds.

> Sleepwalking (somnambulism) does not occur during dreaming, as the muscles in one's core are totally flaccid during REM sleep. Sleepwalking usually occurs during stage III or stage IV sleep. Talking in one's sleep has no relation to REM sleep and can occur at any stage of sleep, but most often occurs right before awakening.

More evidence that the cortex tries to make sense of the incoming stimuli it receives during a dream can be seen from studies that test the effects of external stimuli on people in REM sleep. Many times, but not always, the external stimuli become incorporated into the dream (thus, a loud noise occurring during REM sleep can be incorporated into a dream as an explosion or similar event). The penile erections that occur in males during dreams are often assumed to be associated with sexually explicit dreams; however, even male babies have REM-related erections. Other research has indicated that erections are no more common during sexual dreams than other dreams in males.

The Functions of Sleep

Why do animals sleep? Believe it or not, this is a million-dollar question. For many people the function of sleep is obvious: to get rest. This argument is the essence of one of the major theories on why animals sleep: the *recuperation theory of sleep*. The recuperation theory proposes that being awake for long periods of time disrupts the homeostasis of the body and that sleep restores homeostasis. There are a number of different theories that qualify as recuperation theories, and many single out different functions that sleep restores, such as energy restoration, reparation of the body tissues, and others. According to most of these theories, the period of sleeping naturally ends when homeostasis is restored to some extent.

The other side of the coin is the idea that sleep is not a reaction to the detrimental effects of being awake; it is a result of an internal twenty-four-hour timing mechanism that animals and humans are programmed to follow. According to these *adaptation theories of sleep*, people have evolved to sleep at night because it protects them from predation or misfortunes that could happen in the dark. Thus, sleep has a protective function. These theories tend to focus more on the timing of sleep as opposed to the function of sleep, and some variations of these theories hypothesize that sleep has little restorative function. For these theorists, sleep is like sex in that people are motivated to do it, but it is not essential to individual health.

> There is quite a bit of variation in the amount of time that mammals sleep. A horse may sleep as little as two hours a day, whereas house cats may sleep fourteen hours a day. The giant sloth may sleep as much as twenty hours a day, but that is still not as much as koala bears, which may sleep as much as twenty-two hours a day.

The interesting thing about sleep is that nearly every species of mammal and bird sleeps, and there is evidence that other species sleep, such as reptiles, fish, and insects. This suggests that sleep does serve some important biological function and that adaptation theories of sleep are probably not valid. Moreover, the fact that higher species of animals sleep indicates that the function of sleep, whatever it is, is not restricted to people alone. However, whatever this biological function is has yet to be determined.

There are many puzzling and inconsistent facts about the sleep patterns of animals. There does not seem to be a relation between how many hours a particular species generally sleeps and its size, activity level, or other important features that would indicate restorative functions of sleep. One thing that does seem to make sense is that there is a relation between the daily sleep time of the particular species and its vulnerability to predation. This is more consistent with adaptation theories of sleep. For example, many predators such as lions sleep many hours a day, whereas their prey (often grazers who spend a lot of their time eating) sleep only two to four hours a day. Yet other animals with slow metabolisms like sloths sleep up to twenty hours a day, counterintuitive to the notion that sleep performs a restorative function. So the actual function of sleep remains somewhat of a mystery to researchers.

Sleep Deprivation

One way to ascertain the function of sleep is to look at the research on sleep deprivation. There have been quite a number of studies that have looked at sleep deprivation in human participants. Unfortunately, many of these studies do not control for stress. For instance, most people feel the effects of sleep deprivation during stressful periods in their lives, such as an emergency, work or school issues, changes in schedule, etc. The changes in functioning that people often feel in these instances are confounded by stressful events, as many of the reported changes are similar to the effects of extreme stress. Thus, many of the popular studies of the effects of sleep deprivation are confounded by the effects of stress.

The recuperation theories of sleep would predict that when one is deprived of sleep, this would lead to behavioral and physiological problems, that these problems would worsen as the period of deprivation continues, and that once one does sleep, the amount of missed sleep would be regained.

The results of most of the sleep deprivation research indicates that even mild amounts of sleep deprivation will lead to increases in sleepiness, poorer performance on sustained attention tasks, and some negative cognitive and mood effects. However, the relationship between sleep deprivation and cognitive skills appears to be complex.

First, a substantial amount of sleep deprivation appears to be required to produce consistent alterations of the results on cognitive tests. Secondly, sleep deprivation appears to affect only specific cognitive functions. For example, abilities that are relatively well learned may not be affected. Research investigating the effect of sleep deprivation on logical deduction or on the ability to think critically has also indicated that sleep deprivation does not affect these abilities. However, tests of *executive functions* (e.g., assimilating new information into strategies and plans, switching goals, and more abstract types of thinking) appear to be susceptible to the effects of sleep deprivation.

There has recently been a debate regarding sleep deprivation as a form of torture. Sleep deprivation was used as an interrogation method in wartime settings and recently with terrorist suspects. It is considered by some proponents as a "stress and duress" technique and is legal during interrogations. According to the United Nations, sleep deprivation is a form of torture.

In the overall analysis, the research on the effects of sleep deprivation on one's cognitive abilities has been quite inconsistent and does not fully support the recuperation theories of sleep. Studies investigating the effects of sleep on the immune system have also delivered variable results.

Rebound

A period of sleep deprivation alters the sleep patterns of the sleep deprived person for a short period. Following sleep deprivation, a person's sleep becomes more efficient in that individuals have a higher proportion of stage III and stage IV sleep, both of which are hypothesized to serve the main restorative functions. In addition, people who sleep six or fewer hours per night on a regular basis appear to have the same amount of slow-wave sleep (stages III and IV) as people who get eight hours of sleep per night.

A number of studies have investigated REM-sleep deprivation where only REM-sleep periods are interrupted. During these studies, when an individual begins to transition into REM sleep, she is awakened and then allowed to go back to sleep. When this research is undertaken, the participants typically display what is termed a "REM rebound"; they experience more REM sleep than usual for the first several nights. This tendency of the system of a person deprived of REM sleep to try to compensate for decreases in REM sleep indicates that REM sleep may be regulated differently by the brain than are the other stages of sleep. It is also clear from these studies that REM sleep must be performing some very important functions.

One hypothesized function of REM sleep is the consolidation of memories. However, the research does not always support this. For example, many people who use antidepressant drugs that block REM sleep do not experience memory issues. Other studies have indicated that when deprived of REM sleep, people begin to experience hallucinations and other issues that resemble certain psychiatric disorders. However, these findings are not always replicated.

Disorders of Sleep

Sleep disorders are categorized into either *insomnia* or *hypersomnia*. Insomnia would include any disorder that consists of difficulties falling asleep or staying asleep. Hypersomnia would include any disorder that is characterized by excessive sleeping or sleepiness. Insomnia and hypersomnia are also symptoms of many different psychiatric disorders, such as depression, bipolar disorder, anxiety disorders, and others. Some sleep researchers would include a third class of sleep disorders that consists of any issues related to REM-sleep problems or dysfunctions.

In the general population, about 30 percent of people responding to surveys regarding their sleeping habits report that they have significant sleep-related problems. Many of these reports can be tied to misunderstandings about sleep. For example, many people believe that they must get at least eight hours of sleep a night but naturally function on less than that. Thus, these people are convinced that they have a significant problem with their sleeping habits. Other people may experience problems with sleep related to medication usage or excessive usage of caffeinated beverages, such as coffee, tea, soda, etc. In many of these cases, the anxiety that people feel regarding their "issues" with sleep makes it even more difficult for them to sleep properly.

Disorders of Insomnia

Iatrogenic causes of insomnia are not uncommon (iatrogenic means "created by a physician"). Most often this type of insomnia is a result of medications or the side effects of medications. For instance, the use of sleeping pills, such as benzodiazepines, which are prescribed by a physician, is actually a major cause of insomnia because tolerance to these drugs is developed quite quickly. Once a person develops tolerance, more and more of the drug is required to get the same effect, and patients often begin experiencing withdrawal symptoms of which insomnia is but one.

It is quite common for individuals who claim to have problems sleeping to be overstating their problem. For instance, many people claim it takes them an hour or more to fall asleep, but when studied in a sleep lab, they typically fall asleep within fifteen minutes or less. Years ago when Freudian thought dominated psychiatry, many of these individuals were diagnosed with a neurosis, which is a psychiatric disorder that does not include severe psychosis, such as hallucinations. For many people with insomnia, an effective treatment is *sleep restriction therapy*. Initially, this treatment restricts the amount of time a person can spend in bed, and as time goes on and the person is able to sleep, he is allowed to spend more time in bed as long as he remains sleeping.

Sleep apnea occurs when the person stops breathing multiple times during the night. Each time this happens, the person wakes, breathes, and goes back to sleep. This routine can happen many times at night without the individual being aware of it. Individuals with sleep apnea often complain of being sleepy during the day. There are two types of sleep apnea: (A) *obstructive sleep apnea*, due to some obstruction in the respiratory passages, which often occurs in individuals who snore quite loudly, and (B) *central sleep apnea*, which occurs as a result of the CNS failing to stimulate respiration during sleep. Risk factors for sleep apnea include being male, snoring, being grossly overweight, and increased age.

Restless legs syndrome occurs when an individual experiences twitching or tension in his legs that keeps him from sleeping at night. *Periodic limb movement disorder* involves the involuntary movements of the limbs, often in the legs, which occur during sleep.

Disorders of Hypersomnia

Narcolepsy consists of severe daytime sleepiness that usually occurs at inappropriate times. These people can fall asleep right in the middle of a conversation, while they are driving, and even during sex. People with narcolepsy also experience drop attacks or *cataplexy,* which is a recurrent loss of muscle tone while awake. These attacks may force a person to sit down, or the person may drop as if shot by a gun and remain down but fully conscious. Drop attacks are typically triggered by emotionally charged events. People with narcolepsy may also experience *hypnagogic hallucinations*, which are dreamlike experiences while awake, and *sleep paralysis*, which is the inability to move just as one falls asleep (many people without sleep disorders occasionally experience these). Narcolepsy results from disruptions in the mechanisms that trigger REM sleep and has been linked to reduced levels of a neuropeptide in the cerebral spinal fluid called orexin. Orexin is synthesized in the posterior hypothalamus (which has been linked to states of wakefulness).

Stimulant medications are the most common form of treatment for narcolepsy; however, these often have many side effects. Other treatments that are often used for narcoleptic patients are antidepressant medications, because these medications often suppress REM sleep in people who take them. Other forms of treatment are being investigated.

YOU HAVE TO GROW UP SOMETIME:

The Developing Brain

THE BRAIN BUILDS ITSELF via interactions with the environment and built-in genetic programs. Much assembly is still needed even after birth, as the nervous system requires an incredible amount of assembly and even quite a bit of disassembly as a child matures. As the child grows, some connections are discarded, while others are strengthened. Brain plasticity is the organizing principle of brain development.

Growth and Brain Differentiation

The process of development begins with fertilization and in the creation of the zygote, which results from the joining of the sperm and an ovum. The zygote begins to divide and develop into an organism. The cells differentiate in order to form all of the different systems in the body. Cells must also move to the appropriate areas and group with other like-cells in order to form the structures of the body. In performing this migration, cells also establish relationships with these other cells.

> The term for the zygote after it begins differentiation is the **blastula**, but after about ten days it differentiates into the **gastrula**. The gastrula has three layers: the **endoderm**, which develops into the digestive system, liver, pancreas, and the respiratory system; the **mesoderm**, which develops into bone, muscles, connective tissues, and the circulatory system; and the **ectoderm**, which develops into the skin and nervous system.

The tissue that will eventually become the human nervous system is visible about three weeks after conception on the dorsal side of the developing embryo. This tissue is termed the *neural plate* and initially is on the outermost layer of the embryo. The embryo consists of three layers (working from the inner layer to the outer layer, these are known as the endoderm, the mesoderm, and the ectoderm). Around the time that the neural plate is visible, many of the embryonic cells, which are believed to have the ability to develop into any type of cell if transplanted to the appropriate site (called *pluripotent cells*), become more specified and will only develop into the types of cells in the human nervous system (called *multipotent cells* that can develop into any type of cell of a particular organ). The cells that are in the neural plate are also referred to as embryonic stem cells and appear to have an unlimited capacity for self-renewal and the ability to develop into different types of mature cells. Once the neural plate develops into the neural tube, at about twenty-four days of gestation, the cells become more specified, meaning some will be glial cells while others will be neurons.

At about twenty-one days of gestation, the neural plate folds to become the neural groove, and then the tips of the neural groove fuse together, forming the neural tube. The inside of the neural tube eventually forms into the cerebral ventricles and spinal canal. After about forty days following conception, the forebrain, midbrain, and hindbrain are visible at the anterior portion of the neural tube. After the neural tube is formed, there is a great deal of cell division (called neural proliferation) especially in the ventricular zone. This division is asymmetrical in that the neural stem cells divide such that one cell remains a neural stem cell and the other cell becomes a migratory cell that will move to a different part of the cortex (this is called a migratory precursor cell). The neural stem cell can then continue to proliferate. This process is genetically programmed.

Although there is a great deal of proliferation, the cells are still immature. These immature cells will migrate to their appropriate locations before developing their dendrites and axons. Migration appears to be influenced by the timing of the cell proliferation and their location on the neural tube, with some areas demonstrating more division than others. There are two types of cell migration in the developing neural tube: *radial migration* where the cell migrates in a straight line from the outer wall of the neural tube, and *tangential migration* where the cell migrates in a direction parallel to the walls of the neural tube.

Cells migrate either by *somal translocation*, where an extension grows from the developing cell in the direction in which it is to migrate (this extension elongates and contracts, moving the cell along as it does so), or *glia-mediated migration*. Glia-mediated migration occurs after a great deal of neural proliferation has taken place and there is a temporary network of glial cells, called *radial glial cells*, that appear almost like long ropes for the cells to climb on. These radial glial cells allow the neural cells to move along their structures and to their target site. The pattern of both proliferation and migration is different for different areas of the cortex.

The neural crest forms just behind the neural tube, as cells break away from the neural tube and move dorsally. Cells from the neural crest develop into the neurons and glial cells of the peripheral nervous system, and these travel considerable distances during migration. Neurons are guided in their migration by numerous chemical signals that either attract or repel them.

After the neurons have migrated, they aggregate with other neurons that have migrated to the same area. Aggregation and migration are believed to be facilitated by *cell-adhesion molecules* that are located on the surfaces of the cells. These have the ability to recognize molecules on other cells and adhere to them. It also appears that gap junctions, which are points of connections between adjacent neurons connected by narrow tubes called *connexins,* may also assist in migration and cell aggregation.

The neurons in the appropriate structures then begin to develop their dendrites and axons via *growth cones* that extend and retract fingerlike extensions (filopodia) to determine their correct targets. There are a couple of different ways this may occur. One is possibly a chemical method; the other is called a molecule guidance method. In any event, most of these axons and dendrites find their proper targets. Once this occurs, synapses must be formed, and the most recent evidence suggests that the formation of synapses depends on certain types of glial cells called astrocytes (synaptogenesis).

It is not true that human embryos go through evolutionary developmental stages. This belief was perpetrated by the nineteenth-century biologist Ernst Haeckel, who coined the term "ontogeny recapitulates phylogeny" (development follows the stages of evolution). Haeckel's evolutionary stages were invalid as he used the same embryos but passed them off as different species. Human embryo development contains stages not present in other animals.

Programmed Cell Death

During the process of neural development, many more neurons are created than will be needed, and the system eventually goes through a state of active or programmed cell death called *apoptosis* (necrosis is bad cell death; apoptosis is considered to be necessary). It is possible that some of these developing new neurons are programmed for early death once they have fulfilled their purpose in development. Other neurons may die because they fail to receive stimulation or nutrients. The process of apoptosis and over-proliferation of neurons is believed to be a safeguard within the system, so that if something goes wrong during development and neurons die, there are plenty more that can take over for them. Dead neurons are disposed of by glial cells, and the connections of the existing neuronal tissue fill in the spaces. One of the most important aids in this development is environmental stimulation.

The Old Nature Versus Nurture Debate

The brain continues to develop after birth. So far much of the discussion on neural development has described events that have been driven by genetic programs; however, these cannot function without the interaction of environmental stimulation. A very important feature of development is experience. *Instructive* experiences will direct the development, whereas *permissive* experiences are required for a genetic program to be activated. These experiences are typically time-dependent in that some experiences *must* occur within a particular time interval in order to influence development (*critical periods*), whereas some experiences are more effective if they occur within the particular time interval, but can still have weakened effects outside the time interval (*sensitive periods*). Such periods are specific to the ability involved.

Many textbooks will use the term *critical periods* to describe critical and sensitive periods; however, most of these are really sensitive periods. If neural circuits form, and then are not used, they will not survive or function normally. Therefore, the nature versus nurture debate is really not a debate at all, as neither can work without the interaction of the other.

Vulnerabilities

Toxins and accidents can affect the developing child at any stage. During the first three months following conception, embryos are vulnerable to radiation, viruses, and certain drugs. The period between the fifth and seventh week is when the developing CNS is particularly vulnerable, whereas cardiovascular-sensitive periods occur between the fifth and eighth week. Risk decreases the further into the pregnancy; however, the fetus is vulnerable at all stages of development to drugs and other pathogens.

Perhaps one of the most often cited cases regarding critical and sensitive periods is the case of Genie. From the time Genie was twenty months old, she spent most of her days tied to a toilet in a small, dark room. At night she was placed in a covered crib wearing a straitjacket. Her father punished her if she made any noise at all by beating her and rarely spoke to her. Genie's mother was blind and only spent a few minutes with her every day feeding her. Genie received very little stimulation and grew into adolescence. When she was found, she was thirteen years old, weighed about 62 pounds, and was not quite four-and-a-half-feet tall. She could not stand alone, control her bladder or bowel, or even chew solid food. The severe developmental deprivation led to major problems in trying to teach her language. Her speech was telegraphic (meaning that she didn't use a lot of pronouns and prepositions), and she had trouble with syntax. Her pronunciation was very poor, and she never learned the typical mores of behavior, such as anger control or personal hygiene. Genie's case has been used to highlight the importance of experience in development.

The frontal cortex, especially the prefrontal cortex, displays the most prolonged development of all brain areas, not becoming mature until adolescence. This accounts for the later development of advanced cognitive milestones from age six (grasping basic math concepts) until age fourteen or beyond (developing abstract thought).

Brain Plasticity

The ability of the brain to physically change in response to experience is often termed *plasticity*. Prior to the 1980s the predominant view concerning the development of the brain was that all major developments occurred early and that very few changes took place in the adult brain. Neurons certainly die throughout life, and the view at that time was that adults could not replace them. However, in the early 1980s it was found that certain brain structures in adult songbirds responded with an increase in the number of neurons when the birds learned how to sing. This finding inspired researchers to look at other potential changes in adult brains. Research indicated that adult rats did indeed display new neuron formation in the areas of the hippocampus and olfactory bulbs in response to learning.

Later research indicated that neurogenesis (the growth of new neurons) apparently occurred in the brains of some adult primates as well. Adult rats living in enriched environments displayed nearly two-thirds more new hippocampal neurons compared to rats living in non-enriched environments. It has since been found that the increase in exercise due to the enriched environment was directly related to the increase in new neurons (so it may be possible to slow down the effects of neuronal death by staying active).

When Older Is Not Always Better

Cognitive development may be a lifelong endeavor and may reflect the brain's plasticity. The study of adult neural plasticity is quickly changing the way scientists think about the adult brain. Experiences can increase, decrease, or modify synapses and dendritic spines in adults. However, studies of brain-damaged laboratory animals and clinical studies of brain damage in humans indicate that as people get older, this process begins to slow down. Older adult brains still demonstrate plasticity, but at a slower and more limited pace. This research will be explored further in the section on learning.

The case of Genie and other similar cases highlight the importance of receiving stimulation at sensitive and critical times. Plasticity is a reality in the human brain, but it is restricted by time, as are all things.

THE BIG CHEESE:
The Executive Brain

ACTIONS HAVE REASONS. You pull your hand off the hot stove reflexively to avoid severe tissue damage. But you also turn the stove on, prepare a tasty meal, and organize how a multi-course dinner should be served. Tightly interwoven with attentional abilities are executive abilities, the goal-setting and planning capacities that drive human behavior and distinguish it from almost all other behavior in the animal kingdom. Read on and take a peek at the big cheese, the head honcho, the *you*, the executive brain.

Control

Suppose you decide to have a snack. You stand up and walk to the kitchen to make a sandwich. Several things happened in the sequence: many of them reflexive and many of them more conscious. Standing and walking are activities that you typically do not think about; they become reflexive. Thinking about getting up and thinking about a roast beef sandwich require more conscious mental energy. Walking is a reflexive action; however, you can take direct conscious control of this process. Historically, there has been a debate between the existence of conscious and unconscious mental processes, highly mediated by Freudian thought regarding the unconscious mind. Contemporary views acknowledge that a great deal, if not the majority, of human behavior is performed without conscious awareness. In fact, it is the only way it *can* be performed.

Social cognitive theorists understand the need to explain behavior from a view that includes conscious and unconscious thought. The *Dual Process Theory of Cognition* is a well-established paradigm in cognitive neuroscience that recognizes two modes of processing information: The first one is a controlled mode of processing information that is slower in execution, conscious, and reflective. This is what most people think of when they think of "the mind." It consists of logic, reflection, and reasoning. However, it is slow and can be indecisive. If one had to think "put my foot on the brake" when driving or think out the entire process of braking, accelerating, etc., like a novice driver, there would be many more traffic accidents. In terms of its efficiency, conscious processing is not practical for most daily activities.

The second mode of processing is an automatic mode. This automatic mode is fast in execution and processing, uses less energy, is difficult to change or stop once implemented, and does not conform to rational thinking or logic. A person uses little thought and attentional direction when processing, remembering, and acting in this mode. Automatic processing is driven by schemas and scripts, mental representations or models of the world that are formed from subjective experience.

Schemas are activated during certain contexts, such as evaluating a situation or to determine action or rules for acting. Habits are enacted out of automatic processes, but typically are learned via controlled processes, as in the case of driving a car. Much day-to-day behavior has become automatized, and nearly everything people automatically do was initially a controlled process molded through practice and execution. Many of these patterns are resistant to change; however, nearly every automatic process can be interrupted or changed through conscious effort (this does not apply to purely reflexive movements such as the patellar reflex). Changing automatic processes requires the utilization of energy and attention, and can be difficult.

The extent to which a behavior is "automatic" or "controlled" is not an all-or-nothing situation. There may be some degree of control or automaticity in the same behavior. It is also a mistake to think that controlled behavior requires a "controller" who is totally autonomous. Decisions are often made based on environmental factors or under the influences of motivations (or **drives**).

Executive Functions

Executive functions consist of complex mental processes that allow you to optimize your performance in situations that require the use of a number of cognitive processes or strategies. Executive functions refer to "the big cheese," or the functions that conduct and instruct other areas of the brain to perform, and these also include the capability to inhibit certain processes. Most research associates executive functions with the frontal lobes of the brain. More precisely, executive functions are associated with the prefrontal areas of the frontal lobe and are not tied to one particular cognitive domain (in fact, brain damage anywhere can disrupt executive control). These functions are associated with all thinking and perceiving abilities.

Let's Be Up Front About This: The Frontal Lobes

The prefrontal cortex is simply the most anterior part of the frontal lobes. Most neuro-anatomists divide the surface of the prefrontal cortex into three different sections:

1. *The lateral prefrontal cortex* lies just in front of the pre-motor areas and lies closest to the skull.

2. *The medial prefrontal cortex* lies in between the right and left hemispheres of the brain, in front of the corpus callosum, and anterior to (in front of) the *anterior cingulate cortex*.

3. The *orbitofrontal cortex* (sometimes called the ventromedial prefrontal cortex) is positioned right above the orbits of the eyes and the nasal cavity.

The prefrontal cortex has vast connections with all the sensory systems, motor systems, and areas of the brain involved in memory and emotions. These connections allow the timing and coordination of many different cognitive processes. The lateral prefrontal cortex appears to be associated with sensory inputs; the orbitofrontal cortex appears to be important in the regulation of behaviors; and the medial prefrontal cortex is also important in behavioral regulation, memory, and emotions.

Recall that the primary motor cortex neurons are responsible for voluntary body movements. Something has to inform these neurons when and how they should fire. Setting goals, determining the strategy of a goal-directed plan, and actually executing this plan are functions of the frontal and prefrontal cortex.

The label **prefrontal cortex** can be confusing because the prefrontal cortex comprises about half of the frontal lobes. It is so large in humans because humans have a complex hierarchy of goals, sub-goals, and programs. These flexible representations of goals (programs) are stored in the prefrontal cortex, so it has massive connections within itself and with the rest of the brain.

Complex systems of goals and plans exist at multiple levels in the brain, may be stored as long-term memories or short-term (working) memories, and require numerous connections to implement them. These connections require a well-developed and larger brain compared to animals that respond from a stimulus-response perspective (a stimulus or cue implements a specific plan or response). Humans sometimes respond in such a manner but also have the capacity to rationalize, contemplate, and initiate much more complex and varied responses.

The prefrontal cortex also mediates the consideration of the context of the situation when developing and initiating goal-related behavior. Clinical studies of people with brain damage demonstrate how context and goals interact.

In some cases of people with damage to the prefrontal lobe, there is a *utilization behavior*, where the brain-damaged person will respond by using an available object, even if it is not appropriate to use it. For example, the brain-damaged person may see a fork and spoon in the sink, still dirty, and may pick it up and attempt to eat with it. Another type of problem that can arise due to damage to the prefrontal cortex is called *perseveration*, where the person will continually repeat an action that has already been performed even though the action may no longer be relevant.

Executive functions influence basic cognitive functions, such as attention, memory, and motor skills. Because of this, executive functions are often difficult to assess directly, and many cognitive tests designed to measure other abilities—especially tests that tap complex aspects of attention, memory, verbal abilities, and decision-making—are used to evaluate executive functions.

In order for the type of complex information associated with goals and plans to be stored, it must somehow be represented in the prefrontal cortex of the brain. One must be able to somehow form an image or model of what is happening in the environment, how to implement a goal relevant to the situation, and at the same time, filter and process other stimuli in the environment. The ability to maintain this image or model in the brain is the function of *working memory*.

Do You Have a Plan? The Brain and Planning

Working memory is the technical term for what many people view as short-term memory. It has a limited capacity and a limited duration (therefore it is also considered a component of attention). The classic research indicates that working memory has the capacity of holding five to nine "chunks" of information. A chunk of information is defined as any coherent group of stimuli, such as the letters in a word, a face, a number, etc. The duration of short-term memory is variable but typically is listed at about thirty seconds, although it certainly can be somewhat longer than that and its length varies somewhat from person to person. More modern theories concerning the capacity of working memory indicate that it has different components, such as separate visual and verbal components. Items from working memory can be transferred into longer-term storage as a function of learning.

Chunks of information can be of varying lengths. For example, the numbers 9-2-5 are three chunks of information, whereas 925 is one chunk. A human face with all its complexity is one chunk of information. Chunking like information together can improve executive control over working memory.

The neurons in the lateral prefrontal cortex are responsible for creating a temporary active link between other areas in the brain that code for information to be remembered. Any particular thought is represented in the brain by the activation of a unique set of neurons in the cortex. The prefrontal cortex receives projections from every area in the cortex, and many of these connections are reciprocal. These connections allow information to flow both ways. For instance, you can imagine a hamburger and develop a mental image of it, or you can see a hamburger, close your eyes, and still have an image of it. The activity of the prefrontal cortex neurons results in brain states that are very similar to those states evoked by the object itself. Therefore, one can pursue long-term goals even though the goal may not be physically present. The neurons in the prefrontal cortex become active and remain active while the goal is being processed and/or formed. Once the goal is realized, the brain activation can cease.

A popular model of executive functions is the *Supervisory Attentional System* model (SAS). The key distinction the SAS model makes is between actions that are performed intentionally (those that require some type of attention to be performed) and those that are performed automatically, as discussed earlier in the dual process model. Working memory is utilized during intentional actions. In the SAS model, there are a number of different components.

Familiar actions are stored as schemas. *Schemas* are well-learned plans or templates for action, such as the expected behavior one should engage in while at church. Schemas also consist of certain beliefs and attitudes. A complex task or belief may be stored as a hierarchical collection of schemas, which are often referred to as *scripts*. Schemas and scripts are activated and then utilized by the prefrontal cortex. They are often automatic and result in a swift and defined plan of action.

In the SAS model, the prefrontal cortex is associated with instances that involve making a plan or decision, instances that involve correcting errors or troubleshooting, instances where the response is not well-learned or contains new and unlearned sequences of action, instances that require one to overcome a habitual response, and dangerous or very difficult situations (controlled processing).

Learning from Brain Damage

Often people who have damage to their brains offer clues about brain functioning. For instance, lateral prefrontal cortex damage can result in a tendency to perseverate, the process of sticking to a particular behavior even if it is no longer relevant to the situation. For example, people with severe brain damage may try to tie the laces of their lace-less slippers over and over again. This ability to recognize changes in the "rules" governing a situation is often associated with being flexible and wise. The inability to be flexible in one's behavior can be quite a burden, as the rules governing action in daily situations often change frequently.

While the lateral prefrontal cortex appears to be involved in the ability to consider the context of the situation, some decisions are made based on experience. The orbitofrontal cortex appears to be involved in processing one's life experience and helping to decide whether a choice might be dangerous or inappropriate based on past events. The orbitofrontal cortex is connected to a number of brain structures; the particular structure relevant to connecting experience and action is the *amygdala*. The amygdala is located in the medial temporal lobe area just in front of another very important structure that deals with memory, the *hippocampus*. The amygdala appears to be specialized in learning and in memories of instances of past experiences that have high emotional content associated with them, particularly fear-related events. For example, suppose that when you were young, you were attacked and bitten by a big black dog. You may find that for years afterward, you feel anxious whenever you see a big black dog, even if you don't remember the original incident. This is an example of how the orbitofrontal-amygdala system works. This type of experience or memory is often associated as *intuition*. Intuition has its uses in the sense that it may be a warning based on past experiences; however, intuition can also lead to stereotyping and behavioral rigidity.

Intuition can often prove useful, but it often fails miserably on certain tasks. Automatic cognitive processes like intuition are poorly designed for weighing out decisions that require an understanding of probability or statistics. Good poker players understand this and attempt to learn to ascertain the probability of the cards the other players are holding instead of betting solely on their hand or on hunches.

The orbitofrontal-amygdala circuit communicates its memories with consciousness through "feelings" and not via logic or reasoning. For instance, it has been shown that people with orbitofrontal cortex damage can have difficulties with excessive gambling because they do not weigh their feelings with logic (this is not to say that everyone who gambles excessively or has a gambling problem has brain damage). It is believed that the potential for an occasional high payoff overrides a lack of fear regarding the penalty of engaging in risky behavior when this area of the brain is damaged. The orbitofrontal cortex is also involved in social intelligence, situation-based reasoning, and other vague recollections of earlier experience.

Can You Change That? Correcting Errors

As everyone knows, habits are hard to break. Considerable cognitive resources are required to recognize that certain actions are counterproductive, develop a plan for change, and then implement a plan in the context of behaviors that have become reinforcing and habitual. The function of specific areas of the prefrontal cortex is to be able to recognize errors, both current and potential, override established patterns, and implement correctional action. Individuals who have dysfunctions in these areas of the brain often have difficulties changing their behavior or shifting from one task to another.

The medial prefrontal cortex appears to compare a chosen plan of action with reality. For example, if you are trying to escape from an obnoxious pestering person, and you are at the top of a long staircase, you may feel like jumping, but hopefully you can inhibit this action, because the reality of jumping would likely have very negative consequences. Because the environment in which people live can be unpredictable, it is important to be able to compare goals and plans with what is happening in the real world, evaluate the situation, and make necessary adjustments if need be. There are two areas in the medial prefrontal cortex that are associated with these functions: the anterior cingulate cortex and the posterior cingulate cortex.

Recall that the cingulate cortex is in-between the cerebral hemispheres. The anterior cingulate cortex appears to be involved in the performance of difficult tasks when one is prone to making errors. This area monitors progress toward your goals, corrects problems, and recruits other areas of the brain toward meeting goals. Thus, this area of the brain is activated during difficult or novel tasks, when you make a mistake, or when you must override a habitual response pattern. The anterior cingulate has strong connections with the lateral prefrontal cortex, allowing it to influence working memory with regard to your progress toward goals. Some people with depression, obsessive-compulsive disorder, or schizophrenia have abnormal PET scan profiles in the area of the anterior cingulate cortex that suggest that this area is under-activated. There also appears to be some lateralization in the anterior cingulate cortex such that the depressive disorders are more associated with left anterior cingulate cortex dysfunction, whereas those with anxiety disorders appear to be associated with right anterior cingulate cortex dysfunction.

The posterior cingulate cortex appears more reflective and emotional and may involve taking in social context regarding your goals. The connections of the posterior cingulate cortex are similar to the anterior cingulate cortex.

New Views of Executive Functions

Recent revisions in the SAS model have divided executive functions into smaller modular processes. Some of these models suggest that executive functions are composed of partially independent processes like:

1. The ability to *reject a schema* when you realize that a mental model is inappropriate for the situation

2. *Goal setting*, particular goals are set based on desired outcomes

3. *Control of monitoring and checking*, where you consciously monitor goals and progress toward them, checking outcomes with the desired response and reality

These sub-processes can be grouped into three different stages of planning, such that one stage is devoted to developing a new schema; one stage is devoted to implementing a goal; and one stage is for monitoring progress. These functions may be lateralized such that the right prefrontal cortex may be concerned with monitoring progress, whereas the left prefrontal cortex may be concerned with setting up and developing new schemas.

Models that break down processes into different types of executive functions have been somewhat controversial in cognitive neuroscience; however, studies of individuals with damage to areas of the prefrontal cortex indicate there is some specialization in the prefrontal cortex consistent with these models.

Not Thinking?

Just because many actions are performed without conscious awareness does not mean that a person is not thinking. Likewise, only a small percentage of all brain activity actually reaches conscious awareness, but that does not mean that the brain is not processing these stimuli. People with extensive damage to their primary visual cortex do not respond to visual stimuli that are projected to the damaged areas. In clinical experiments, these individuals claim that they cannot see anything in these visual fields; however, when pressed to guess at what was presented to them, they consistently guess correctly at a significantly higher rate than expected by chance. These experiments indicate that while these patients are unaware of the stimuli in their visual fields, it is quite possible that some processing in the brain is still occurring. This phenomenon has been called *blindsight*. It is unclear whether the conscious processing in these cases occurs in the subcortical areas associated with vision, or whether there is leftover processing being accomplished in areas of the cortex, but some form of processing stimuli in these damaged visual fields is still occurring.

Where Has Free Will Gone?

In a much-cited series of experiments, Benjamin Libet asked participants to sit calmly and watch a special clock, and whenever they decided to move their hand, they were to notice the position of the clock hand (recording the time). The participants were also monitored by EEGs recording their brain's electrical activities. Libet compared the EEG recordings to the time the participants noted that they decided to move. What was typically found in these cases was that the EEG recorded brain activity in the appropriate area of the motor cortex about one-half second before the participant's awareness of their desire to move, which is called the *readiness potential*. The idea here is that the brain appears to be active before you are consciously aware of your actions. This particular phenomenon has been used to denounce notions of free will and to claim that free will is an epiphenomenon (a secondary or additional phenomenon; a byproduct). Libet's and others' experiments resulted in quite a debate regarding free will in human behavior. However, there are a number of issues with this.

The first issue goes back to the notion of the dual processing theory and that movement is an automatic process that operates outside of conscious awareness. Philosophically, the choice was made when the participant chose to be part of the experiment. Once that happened, a different process took over that is more efficient. More recent replications of this study have indicated that the readiness potential is not as Libet envisioned it and probably represents a slow buildup of electrical activity prior to spontaneous movement, which reflects the goal-directed neural operations that lead to these movements.

> Interestingly, society is based on the principle of autonomous behavior. One would be hard-pressed in a court of law to explain the reason for an infraction as "My brain did it!" It is doubtful that, in the absence of any brain damage or psychiatric disorder, one could blame his brain for his behavior and absolve himself of legal responsibility for his actions.

Where is free will in the brain? That is a good question, because damaging any of several different brain areas can result in the loss of the capacity to make decisions. Moreover, extensive damage to the brain that occurs in diseases like Alzheimer's disease or other neurological conditions like strokes can also leave one incapable of making informed choices. The issue of "what is free choice?" is a complicated philosophical question muddled by a number of internal and external variables. For example, at one end of the extreme, some argue that people have total free choice in everything they do; however, this is obviously not the case. People are restricted by their own physical limitations, the environment, and their station in life as to what their choices are. On the other hand, some say that people have no free choice and that everything is dictated by genetics and/ or the environment. This point of view is equally untenable as people always have control over how they react to their limitations and opportunities. Probably the best answer to the question of where free will is located in the brain is that volitional choice is dependent on a number of brain structures working together, and that disrupting any particular one of them, or several of them, can lead to limitations in the ability to make choices.

SAY THAT AGAIN?
Attention and the Brain

WHEN SOMEONE SAYS "PAY ATTENTION," what exactly does that mean? Attention consists of a number of different processes by which people select certain information for further processing, while discarding other irrelevant information. It would be impossible for you to process all the information received by your brain; this would result in a sensory overload. How do you decide what to pay attention to? This chapter deals with the different types of attention, how they function, and what happens when certain attentional processes are disrupted.

The Anatomy of Attention

In the 1890s the American psychologist William James proposed that attention was composed of two aspects: "reflexive" (in modern terms automatic processes) and "voluntary" processes (controlled processes). Other characteristics of attention have been identified, such as the capacity to disengage from one stimulus and to shift focus to another or the ability to remain focused and ignore distractions. Most researchers conceive attention as a system of processes where there is sequential processing of stimuli in a series of stages involving different systems of the brain. This process is organized hierarchically in that the early stages are governed by the brain areas specific to the modality in which stimuli are processed (i.e., visual, auditory, etc.), and then later processing involves the cooperation and integration of different brain areas.

A characteristic of human attention is that it is limited in its capacity. The brain can only process so much activity at one time, so for example, you may be unable to concentrate on a conversation and perform a calculus problem simultaneously. Attentional capacity varies between individuals and within the same person over different times or tasks. There are several other important aspects regarding the anatomy of attention:

- The immediate *span of attention* is how much information you can grasp at one time. This can range from being relatively effortless to effortful and draining. It is surprisingly consistent across species and appears to be resistant to aging. The span of attention is a component of working memory and is limited in that no one can grasp the aspects of all the stimuli bombarding him or her at any given moment.

- *Selective or focused attention* (concentration) is the aspect of attention that most people refer to when using the word "attention." Focused attention is the ability to target a small number of stimuli or thoughts (most often one or two at a time) while filtering out other distracting stimuli or thoughts.

- *Sustained attention* (vigilance) is the ability to maintain your concentration over a particular length of time. For instance, reading this entire chapter and ignoring distractions would be an example of sustained attention.

- *Divided attention* is the capacity to attend to multiple stimuli that compose a complex task, such as driving a car in heavy traffic, or the ability to attend to more than one different task at a time. Divided attention typically results in a drain on your overall attentional capacity.

- *Shifting or alternating attention* is the capacity to disengage from one task and focus on another task.

Some of these aspects of attention work together, such as selective and sustained attention, whereas others may be opposed to one another, such as divided and shifting attention. All of these aspects of attention are mediated by working memory and therefore heavily dependent on the frontal and prefrontal cortex of the brain. However, because these processes are so complicated and involve other areas of the brain, damage to other areas of the brain can disrupt attention. In fact, any stimulus or event that places an additional burden on the system can result in difficulties with attention. Moreover, any disruption involving a single component of the attentional system can lead to difficulties in more than one of these aspects of attention.

> Attention is among the most sensitive of cognitive abilities. Difficulties with attention and concentration are among the most common cognitive problems associated with psychiatric disorders such as depression or anxiety and are typically the results of brain damage to almost any area of the brain.

One of the biggest culprits contributing to difficulties with attention is a decreased rate of processing information (how fast you can process information). A decreased rate of processing can result from any number of sources, including brain damage, aging effects, distractions, anxiety, the effects of medications, psychiatric problems, fatigue, hunger, stress, etc.

The Prestriate Area of the Brain and Attention

Theories of selective attention were traditionally divided between theories of early levels of stimulus selection and theories of late selection. The early selection theories purported that there was a filtering mechanism by which "channels" of irrelevant stimulus input would be ignored or rejected for further processing based on a physical attribute of the stimulus. Conversely, late selection purported that all stimuli are processed in the same considerable detail before any selection of relevant versus irrelevant stimuli via attention takes place.

Much of the current research regarding the neuroanatomy of selective visual attention has focused on the *prestriate* area of the brain. The prestriate is the area directly anterior to the primary visual cortex (or striate cortex). Functional brain imaging in both humans and primates has determined that there are two pathways in the prestriate cortex that are important in attending to visual stimuli: a ventral stream (bottom or belly side that passes through the temporal lobe) and a dorsal stream (top or backside that connects to the parietal lobe). It appears that paying attention to shape, facial identity, or color produces more activation in the ventral stream (a "what" pathway), whereas paying attention to movement or position is associated with greater activity in the dorsal stream (a "where" pathway; see **Figure 3-3**).

Most of the studies of auditory attention focused on the effects of attention on the processing of sounds in auditory cortical areas, and less work has focused on the neural structures and mechanisms that direct and control auditory attention. There appear to be two similar streams of selective attention for auditory information as well (a "what" stream and a "where" stream; see **Figure 4-3**).

It would be difficult to pay attention to an item if your brain does not first process the sensory information associated with the specific environmental stimulus. It is important to understand that paying attention to visual information requires that first this information be processed in the areas associated with vision; likewise, stimuli from other sensory modalities must first be processed via their specific brain mechanisms.

Working Memory: Who Does It Work For?

Because the process of attention requires a focused effort to control your cognitive processes, attention is strongly tied to executive functions and the frontal area of the brain. Recall that working memory is a capacity-limited, duration-limited function of the brain heavily associated with the prefrontal cortex. You can think of working memory as the modern theoretical interpretation of what used to be termed *short-term memory*. Working memory is that component of attention and memory that keeps needed information online and available for use for a very short period of time. One of the most influential models of working memory is the model by Baddeley and Hitch that purports that working memory is made up of a *central executive* responsible for controlling and coordinating the operations of two specific subsystems: *a phonological loop* and a *visuospatial sketchpad*:

1. The central executive is responsible for running the system of working memory by allocating information to the two subsystems. The central executive is also involved in cognitive tasks, such as performing mental arithmetic and other types of mental problem solving that require mental attention.

2. The visuospatial sketchpad is responsible for storing and processing visual or spatial information for short periods. This system is used for a number of immediate functions, navigating through the environment being among them.

3. The phonological loop deals with auditory and written material. It is the part of working memory engaged when trying to retain a new phone number for short periods. It consists of two parts: (1) the phonological store that is linked to the perception of speech and is able to hold speech-based information for very short periods, and (2) the articulatory control process linked to the production of speech and used for the rehearsal and storage of verbal information from the phonological store.

Why Can't You Pay Attention?

The term *attentional control* is the technical term for your capacity to select what specific stimuli in the environment you will attend to and what you will not pay attention to. Most people would equate attentional control with their ability to concentrate. The neuroimaging research has indicated that attentional control is mediated by the anterior portions of the brain, including the prefrontal cortex and the anterior cingulate cortex. Concentration, or attentional control, is related to executive functions and to working memory, and to the executive control component of working memory.

Early researchers proposed that there were three aspects of attention involved in the ability to maintain focus: alertness (remaining aware), orientation (receiving information from the senses), and executive control (which combines volition and automatic processes). These aspects of concentration are quite variable among individuals and demonstrate a wide range of variation within the same individual. All other things being equal, the ability to concentrate on something is strongly linked to your motivation to do so. For instance, a student reading what she considers to be a boring book may experience difficulty maintaining her focus, whereas a different student reading the same book, but who considers it interesting, will experience less difficulty.

> Mild stimulants, such as caffeine, at low doses have been demonstrated to increase selective attention in some studies and to interfere with attention in other studies. There appear to be a number of interacting variables, such as gender effects, age effects, tolerance, and of course, dosage effects.

One mistake people make about attention and concentration is the notion that you should be able to easily focus on anything. Barring any mental or neurological disorders, maintaining attention on something is really a matter of motivation, eliminating distractions, and setting goals.

The thing to remember is that selective attention is strongly tied to working memory and that working memory consists of a number of brain processes that have a limited capacity and a limited duration. When engaging working memory, you are using cognitive resources that tax the system. Your ability to focus in the presence of numerous distracting stimuli should not be taken to mean that anyone could do so. It is important to understand your own capabilities and to structure your environment around those. This may include eliminating potential distracting sources and setting goals, such as reading or studying in short sessions with frequent breaks.

Another thing to keep in mind is that the ability to focus or sustain attention is like a muscle; the more you use it, the stronger it gets. However, the benefits of sustaining attention on one task may not always generalize to the ability to sustain attention on other tasks. For example, a star baseball pitcher may be able to focus on the batter and ignore the crowd in the stands and yet not be able to read a book on a crowded bus. One of the reasons for this may have to do with motivation and personal preference; another may be related to a lack of understanding of how to put oneself in the same "mindset" across different tasks.

Disorders of Attention

Any number of neurological or psychiatric conditions are associated with disrupted attentional abilities. This section will focus on some of the better-known disorders that are specifically disorders of attention.

Attention Deficit Hyperactivity Disorder (ADHD)

There are many misconceptions about ADHD. First, ADHD is a clinical disorder that must have been present during childhood. Despite what you may have read to the contrary, there is no such thing as adult ADHD (at least not yet). If an adult is diagnosed with ADHD, there must be evidence for the presence of the disorder in that person's childhood history. Adults certainly can develop attentional disorders, but often these are due to the onset of some psychiatric disorder or neurological problem and are technically not ADHD.

Secondly, since ADHD is a disorder, it cannot be commonplace. ADHD is not due to a virus or germ, as is the case with the common cold, and represents a disorder of normal function (broadly defined as a range of abilities that the majority of children possess).

Additionally, there are no medical tests that can identify the presence or absence of ADHD in a person, such as a blood test or particular brainwave pattern. This is the case with the majority of psychiatric disorders. The diagnostic criteria for ADHD were decided on by a committee, and the diagnosis of ADHD is made on the basis of behavioral signs and childhood history.

Finally, the cause of ADHD is unknown, and speculations regarding the use of sugar or food preservatives or other causes have never been verified. Diet can certainly affect a person's attentional capacities, especially those of a young child, but diet alone does not appear to be a cause of ADHD. However, one certainly cannot rule out the effects of a very poor diet on brain development in a child. But since ADHD affects children of all socioeconomic statuses, it appears that diet alone cannot be a specific cause of ADHD.

> There is technically no such thing as ADD. There are three different types of ADHD: a primarily hyperactive type, a primarily inattentive type, and a type that is a combination of these (hyperactive and inattentive). The primarily inattentive type of ADHD is what many people wrongly refer to as ADD.

ADHD appears to be related to the frontal and prefrontal cortex in children. One of the most popular theories regarding the mechanism of ADHD was the notion that the normal or relaxed rate of neuronal firing in these children was for some reason slower than that of children who did not have ADHD. This led to the use of stimulant medications such as Ritalin to paradoxically treat the inattention and hyperactivity in these children. The thought was that since these children were naturally under-stimulated, the shifting attention and hyperactivity was a form of self-medication to get their systems going.

The use of stimulant medications, such as Ritalin, has been successful in treating ADHD in many cases, but there are a number of side effects that are associated with these medications, and the medication does not always work. Moreover, stimulant medications are more effective for children that have a hyperactive component to their disorder and are not as effective for those that only have the attentional component. Recently, there have been other medications that are not stimulants that have been effective to some extent in treating ADHD. In addition to medications, behavioral or environmental changes can also help these children.

Hemispatial Neglect

Patients with hemispatial neglect typically will not pay attention to objects in their left visual fields. These individuals have typically experienced damage to one side of their brain and fail to attend to the stimuli on the opposite side of the brain that has the lesion. The damage that occurred is often the result of a stroke, and typically this occurs on the right side of the brain (thus resulting in neglect of objects in the left visual field). Right-side neglect is rare as there are redundant connections for the right visual field via both the left and right hemispheres, whereas attention to the left visual field appears to be a function of the right hemisphere only.

Hemispatial neglect appears to be a disorder of degree such that varying levels of damage produce different levels of neglect for the left side. Some patients may just neglect the left side of objects in their visual field, and others may neglect the whole left visual field. One study examined the brain regions of patients who had neglect and concluded that the critical region that was damaged in patients with left neglect was the right angular gyrus in the right parietal lobe. Other studies have suggested that damage to the right superior temporal gyrus is associated with a less severe form of neglect.

Neglect patients typically do not attend to objects in their left visual fields. When these patients are asked to draw an object, they will often omit the left side of the object or try to cram everything into the right side of the drawing. Some studies have indicated that neglect is not just limited to vision but can also include auditory and tactile stimuli.

Balint's Syndrome

These patients have extensive damage to both the right and left parietal lobes resulting in severe disturbances when dealing with spatial information. In addition to having difficulties using visual information to guide their motor movements (a condition known as optic ataxia) and a fixed gaze without any deficits in eye movement (known as optic apraxia), they also experience the inability to perceive more than one object at a time (called *simultanagnosia*). The patient may notice a chair in a room, and then all of a sudden the chair disappears and the patient sees who is sitting in the chair. Typically these patients have normal vision and can localize sounds, but for some reason, they are not able to pay attention to multiple objects at the same time.

WHO WAS THE FIFTH PRESIDENT?

Learning and Memory

PROBABLY THE MOST STUDIED area of cognition is the area of learning and memory. For most people their memories define who they are, who they want to be, and who they will be. Learning and memory are actually two ways of thinking about the same phenomenon, in that they both describe an outcome based on the ability of the brain to change its structure and its functioning as a consequence of experience. This chapter investigates the major processes in brain structures involved in the process of learning and memory.

Learning

Technically, learning is described as a change in brain structure in response to experience, and memory is associated with how these changes to the brain are subsequently stored, represented, and reactivated when needed. Someone unable to learn or remember her experience would find every single moment as an original experience.

Learning new material or changing material already learned requires important changes in the structure of the brain. For the most part, the *plasticity* of the brain, or the ability of the brain to change in response to experience, is driven by genetic programs. In some animals specific types of learning occur via genetic programs; however, animals can also learn from experience. Certain animals, such as certain types of birds, immediately learn to associate themselves with the first living object they see upon exiting their shells as hatchlings, and will identify with that animal. This process is called *imprinting* and has survival value, as typically the first thing these animals see is their mother. Imprinting has been taken advantage of by some working in the pet trade where breeders of exotic birds, such as parrots, make sure that they are present when the chicks hatch and then rear and feed the young bird by hand. When this is done, the developing parrot believes itself to be a person and makes a better pet.

> Classical conditioning occurs in a number of contexts, such as feeling nervous before going to the dentist or flinching before an expected loud noise. Neuroimaging techniques have implicated right hemisphere activation, and especially right frontal cortex activation and cerebellar activation, as being associated with learning via classical conditioning.

Imprinting does not occur in people, but several types of learning have been identified. Some of the best-known early learning experiments with animals that were applied to human learning came from the Russian physiologist Ivan Pavlov, who was studying digestion in dogs. Pavlov noticed that dogs would salivate, a reflexive action in dogs when food is present, before they were presented with any food. He performed a series of experiments and learned that the dogs had associated the presence of the white laboratory coat worn by the laboratory assistants with a soon-to-be presented food object. This type of learning, where a reflexive action is associated with an environmental stimulus, has been termed *classical conditioning*. Conditioning is a term used in psychological learning research to imply a specific type of learning that involves the association of two or more stimuli. In the classical conditioning model, the learning takes place via the association of a particular reflexive action (i.e., salivating) with some environmental stimulus that normally would not elicit this response (i.e., a white lab coat).

It works like this: A stimulus that elicits the reflexive action, such as the presence of food leading to salivation (the food stimulus is called the *unconditioned stimulus* [UCS] and the reflexive action of salivation is termed the *unconditioned response* [UCR]), is paired with some stimulus that normally would not elicit the reflex, such as a white laboratory jacket (this new stimulus is termed the *conditioned stimulus* [CS]). Over time, the animal learns to associate the white laboratory jacket with the presence of food and salivates whenever he sees the lab coat (this new response to the lab coat is termed the *conditioned response* [CR]). Pavlov initially believed that the conditioned, or learned, response was identical to the original unconditioned response; however, the conditioned response is actually a bit weaker. In addition, later experiments indicated that classical conditioning cannot pair any unconditioned stimulus to an unconditioned response (for example, it would be nearly impossible to get dogs to learn to salivate to a severe electric shock). Pavlov, who won a Nobel Prize for his work, is considered one of the pioneers of learning and memory.

Operant Conditioning

While classical conditioning was being studied in Europe, a different type of learning or conditioning was being studied in the United States. Pioneers Edward Thorndike and John Watson investigated what later came to be termed *operant conditioning*. Probably the best-known proponent of operant conditioning is the behavioral psychologist B. F. Skinner. Operant conditioning works on the assumption that a particular behavior will either increase or decrease in response to being reinforced or punished. A *reinforcer* is anything that increases the probability of repeating a behavior, whereas a *punishment* decreases the probability that the behavior will reoccur. For example, a pigeon can be taught to peck a lever in response to a red light if every time he does so, he is provided with a food pellet immediately afterward. Soon the presence of a red light will emit the pecking of a lever in the pigeon. In this type of learning, many different behaviors can be learned if reinforced. Skinner became interested in how scheduling patterns of reinforcement (or what many people would call rewards) affected learning acquisition and the stability of the newly learned behavior. He found that learning occurred more quickly if continually reinforced, but once learned, a behavior could be maintained longer (the behavior was stronger) if a variable schedule of reinforcement was used.

Operant reinforcers are used extensively in society: grades for school performance, medals and trophies for athletics, and money for labor. Money represents a special form of reinforcer called a token that has no intrinsic value in itself, but represents a means to procure other tangible rewards.

Learning Without Reinforcement?

Some theorists did not believe that learning occurred only via reinforcement or punishment. One of these people was the cognitive psychologist Edward Tolman. Tolman allowed rats, who were neither hungry nor thirsty, to wander in a maze. This particular maze was in the shape of a T. At one end of the T, Tolman had placed water, and at the other end of the T he placed food. Since the rats were neither hungry nor thirsty, they had no motivation to go to either end of the T and simply explored the entire maze. Later, Tolman withheld food from half of the rats and water from the other half. When he rereleased them into the maze, the hungry rats went right to the food and the thirsty rats went right to the water, a finding that is inconsistent with what operant conditioning would predict. The rats learned without reinforcement. Tolman hypothesized that the rats had made cognitive maps of the maze and did not need to be reinforced to learn where the food and water were. Tolman's study was followed with other research and would actually later develop into the field of cognitive psychology and cognitive neuroscience.

Albert Bandura performed another famous series of experiments that indicated that learning does not have to be directly reinforced. In these experiments, children watched a person being rewarded or reprimanded for beating up a clown doll (called a Bobo doll). When the children watched someone being rewarded for punching out the hapless doll, they themselves were later more likely to repeat that behavior, whereas children seeing someone punished or reprimanded for beating up the doll were significantly less likely to beat up Bobo. Bandura termed this form of learning "modeling," in which the mere observation of seeing someone perform a task or behavior can lead to learning.

Of course there have been many modifications to these theories of learning, but learning can occur by any of these particular methods. Moreover, humans often learn for the sake of learning or for non-tangible rewards and other long-term potential reinforcers, as opposed to animal models where delays in reinforcement often lead to slower learning. While a certain amount of learning does occur via classical and operant conditioning in humans, there is a cognitive component that is associated with learning, and learning in humans often occurs via a combination of these previous methods.

The brain must somehow represent the learning process. Human and animal brains are plastic in that experience can directly lead to physiological changes in the brain, which can result in learning. In order for learning to be successful, there must be some structural change in the brain that occurs as a result of learning and allows the organism to store the memory of the newly learned material. If this were not the case, you would have to learn the same thing over and over again. There has been quite a deal of research investigating how the brain changes during learning. One of the most salient findings as a result of this research is the discovery of long term potentiation in the brain.

Long-Term Potentiation

Recall that most neurons communicate via a neurotransmitter sent across the synapse from the axon of a sending neuron to the dendrite of another neuron. The sending neuron is termed the presynaptic neuron and the receiving neuron is often termed the postsynaptic neuron. Learning at the cellular level would be dependent on an increase in the probability that a postsynaptic neuron would be activated in response to the neurotransmitters released from a presynaptic neuron. The more quickly this response takes place, the more learning is accomplished. Experimental studies have attempted to replicate this process by briefly applying high-frequency stimulation to presynaptic neurons and observing the response in the postsynaptic neuron. When this experiment is performed on the brains of rats and other animals, changes occur in the postsynaptic neuron that include the sprouting of new dendritic spines and a faster response to messages from the presynaptic neuron. When more presynaptic neurons that all synapse (connect) with the same postsynaptic neuron are stimulated, the effect is even stronger. This effect has been termed long-term potentiation (LTP).

LTP requires that inputs are excitatory. LTP occurs in the hippocampus and associated areas, such as the entorhinal and parahippocampal cortices, which use the major excitatory neurotransmitter glutamate. Typically many different inputs synapse on the neurons in these areas, and LTP is believed to occur in these areas when new memories are formed.

LTP has two characteristics: First, LTP can last for quite a long time after it is established (even several months after the initial stimulation), and secondly, it only occurs when the firing of the presynaptic neuron is followed by the firing of the postsynaptic neuron. The changes that occur as a result of this relationship between the presynaptic and postsynaptic neuron are further evidence of plasticity in the brain. Other findings indicate that LTP is most prominent in brain structures that are associated with learning and memory.

Types of Memory

One of the most influential models of memory, proposed by Atkinson and Shiffrin in 1968, has been termed the *modal model of memory*. The following figure depicts this model:

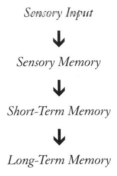

In this model, memory follows a linear process starting first with the sensory storage stage. This stage has a very brief duration (only a few seconds at most), a very large capacity, and has different representations for each sensory modality. For example, if you glance at the room around you and then close your eyes, there will be a very short period of time (less than a few seconds) where everything in your span of vision is stored in your visual sensory memory like a snapshot. The image will quickly dissipate.

Attending and Rehearsing

In order to retain something from sensory memory, you must focus on it or pay attention to it. However, you can only focus on various aspects of sensory memory, and much information is lost in a short time. Attending to some event or stimulus from sensory storage allows you to transfer this information into the second mode, short-term memory, which has a very short duration (about thirty seconds or less) and a very limited capacity (five to nine items). This is demonstrated by the fact that if you try to remember the entire scene, you will only be able to focus on a very small portion of it at a time.

If you want to remember information for a longer period of time, you need to either practice or rehearse it. If you do this efficiently, it transfers into long-term memory, which according to this theory is essentially of unlimited duration and unlimited capacity.

Of course, the understanding that short-term memory is much more complicated than this model suggests resulted in the idea of working memory. Nonetheless, this Atkinson-Shiffrin model has endured over time with some adjustments (e.g., the replacement of short term memory with a more complicated notion of working memory and the division of long-term memory into separate subtypes). This particular model has also led to the development of the idea of how a transfer from one particular stage to another memory stage occurs, and to the notion of being able to retrieve memories from long-term storage (which many equate with the process of memory). More current conceptualizations of this model include a reworking of the short-term memory component that includes both sensory storage and working memory.

Why do you always forget where your glasses are, but never forget how to read?
Forgetting where you put your keys or glasses are typically examples of poor encoding: Your attentional resources are divided, and you do not sufficiently encode your actions. Forgetting how to read is more difficult, as reading is a well-practiced procedural task. However, significant brain damage can lead to an acquired reading disability (alexia). Most reading disabilities represent learning disorders.

The modal model of memory has influenced the process model of how people learn and remember information. In this sequence of events, one is exposed to some environmental stimulation (sensory memory), pays attention to it (thus putting it in working memory, a process commonly called *encoding*), and then rehearses or practices it (thus transferring it into long-term storage, a process known as *consolidation and storage*). When one wishes to pull this memory from long-term storage, she engages in a process known as *retrieval* or *recall*. Long-term potentiation refers to the physical process involved in consolidation and storage of memories. However, all memories are not equal in long-term memory. The component of memory that most people identify as their *real* memory has been divided into several different types of memory processes.

Divisions of Long-Term Memory

The first major distinction that has been made in long-term memory is whether the memories in this type of storage are consciously accessible; whether there is a need for conscious recollection or recall when reproducing these memories (in other words, these memories would be part of the controlled processes in the dual process model of cognition). Memories that are consciously accessible have been termed *declarative* or *explicit memory*. Memories that are not consciously accessible have been termed *non-declarative memory* or *implicit memory* (this division would consist of automatic processes according to the dual process model of cognition). Declarative and non-declarative memory are also divided into several subdivisions.

Declarative memory can be further divided into memories for events that are personal (*episodic memory*) and memories for factual information (*semantic memory*). Personal events include your recall of when you graduate from high school, your first boyfriend or girlfriend, etc. Factual events include such things as who was the fifth president of the United States, what to use a screwdriver for, what the meaning of the word *memory* is, etc.

> While there is quite a bit of supporting evidence for these different memory systems, there is also some overlap between them. In addition, the brain structures associated with the consolidation and retrieval of these particular memories differ depending on the type of memory involved. Declarative memories are formed in the hippocampal structures and sent to association areas.

Non-declarative memory has several different divisions: *procedural memory* (cognitive and motor skills), *perceptual representations* (systems for perceiving words, sounds, objects, etc.), *conditioned responses* (this can include both classical and operant), and *non-associative learning*. Non-associative learning consists of a number of processes, such as habituation, where one ignores the continual presence of a stimulus (not hearing background conversations when concentrating on something else), and sensitization, which is the opposite of habituation (becoming more sensitive to stimulation that is particularly relevant or threatening). While both declarative and non-declarative memory are important in functioning, traditionally researchers have focused on declarative memory, as it occupies such an important focus in the lives of most people and is much easier to study than forms of non-declarative memory.

The Biological Foundations of Memory

Working memory requires the use of the prefrontal cortex and the hippocampus. Much of the information that enters and is used in working memory is disposable; it does not need to be saved over the long-term. The neurons in the areas of the prefrontal cortex maintain their firing rates, representing various sensory inputs so that you can use working memory to perform a specific task, such as ordering off a menu, dialing a phone number, following directions, etc. Some of this information may enter long-term memory depending on how relevant it is to you. Much of this information may require some form of additional processing, such as rehearsal, to transfer it into long-term memory. The process of transferring information from short-term, or working, memory into long-term memory requires the action of the hippocampus (although there are types of information that do not appear to be dependent on the hippocampus to enter long-term memory).

The Hippocampus and Memory

The most famous clinical case regarding the memory and the hippocampus is the case of HM. HM had both his temporal lobes removed due to intractable epilepsy, thus removing both hippocampi. Following his surgery, he scored relatively normal on IQ tests, but could not form any new memories. For instance, when his mother died, he could never encode this information. Years later, even after being told many times, if someone mentioned that his mother had died, he would break down and cry as if he were hearing this information for the first time. However, HM could learn some new procedural information like figuring out how to get to the lunchroom in the clinic he was in, even though he could never remember how he learned how to get there or having been there before.

The hippocampus is in the medial area of the temporal lobe and receives inputs from nearly every area of the cortex. The hippocampus is part of a larger system known as the limbic system that is important in processing emotions and in memory. The limbic system includes the amygdala, the orbitofrontal cortex, and the anterior cingulate cortex. The hippocampus has specialized, adjustable synaptic glutamate receptors known as NMDA receptors, named after their primary neurotransmitter (N-methyl-D-aspartate), which are able to define objects and collections of objects within particular contexts. The hippocampus is composed of a set of modifiable neurons that receive input from the cortex and represent what is going on in the current environment. For instance, the hippocampus right now is representing that you are reading this book, the phrases you are reading right now, where you are sitting or standing, and other events currently happening within this context. The hippocampus receives projections from various areas of the cortex that represent these particular aspects of your immediate experience. These hippocampal connections are strengthened via long-term potentiation in the cortex.

In addition, the connections between the hippocampus and the cortex are reciprocal. This allows the connections that have been activated in the hippocampus to be sent back to the cortex from the hippocampus. So if the process of forming a memory is involved in the actual representation of the experience itself in the cortex, then this experience is sent to the hippocampus and a recreation of the experience is sent from the hippocampus back to the cortex where the memories become stored in the association areas. When a person continues to rehearse information, this activity reverberates back and forth between the cortex and the hippocampus, modifying the neurons in the cortex itself so that it can reproduce the activity that was originally associated with a person's experience. It is believed that long-term memories that are developed in the hippocampus and sent to the cortex are stored or maintained in the same cortical areas that represented the initial experience. The hippocampus also interacts with the prefrontal cortex that houses working memory to maintain memories long enough for you to use them and transfer them back to the cortex.

REM and Memory

Another aspect of transferring information into long-term memory was discovered by means of training rats how to navigate through mazes. It was determined that when the rats slept after the training sessions and entered REM sleep, their hippocampi continued to play back and forth the correct solutions to the mazes. This playback occurred several times during the night (this was determined by the patterns of activation in the rats' hippocampi during REM sleep compared to the patterns of activation during running and learning the mazes). If the rats were prevented from entering REM sleep, they experienced difficulty consolidating the memories regarding learning the maze. Therefore, it is believed that if an activity is rehearsed enough, much of the consolidation from short-term, or working, memory into long-term memory takes place during REM sleep, although not all the evidence supports this.

Despite popular opinion, memory is not like a tape recorder. When you recall declarative memories, your brain must actually recreate them. This recreation is subject to outside influences (context, suggestion, emotional states, etc.) and can alter the memory of past events. In psychology, it has been long known that outside influences can change the recollection of past events so as to distort them significantly.

The Frontal Lobes and Memory

One question that neuroscientists were concerned about was the contribution of the frontal lobes to the memory process. As mentioned previously, the frontal lobes send projections to the hippocampus, but the information from the frontal lobes may not consist of content information, such as the identity or meaning of something. It appears that the frontal lobe inputs add information regarding context and information about a particular event or environment, rather than just semantics. The hippocampal circuits also appear to be involved primarily in declarative memory and not in non-declarative memory. The famous amnestic patient HM discussed earlier was unable to form any new declarative memories; however, he still was able to learn procedural information (non-declarative memory) despite being unable to recall ever having performed these tasks before.

Procedural memories appear to be formed in the basal ganglia, stratum, and cerebellum. It may well be that certain types of classical conditioning are represented or formed in the cerebellum and in the limbic system.

Forgetting

Most people believe that forgetting is bad; however, it may be more of an efficient use of memory capacity as opposed to a fault of the system. The availability to previously stored information needs to be prioritized in the system so that the most relevant information can be accessed easily. In addition, as any student will tell you, the ability to recognize information is easier than the ability to recall information (this is why students prefer multiple-choice over essay examinations). The amount of processing at the encoding stage that free recall requires is much greater than what is needed for recognition. Some researchers have proposed that recognition consists of two separate mechanisms: *familiarity*, where the item just feels familiar and is context-free, and *recollection*, where context is important and cues memory. These two mechanisms of recognition may also explain why it is easier to recognize things than to recall them.

The way that information is processed will also determine its availability to be retrieved. The *levels-of-processing account theory* proposes that information processed in a semantic fashion will be easier to retrieve than information processed in a perceptual fashion. Research has suggested that the frontal cortex may be important for selecting attributes that allow for easier and quicker retrieval of material based on studies using neuroimaging. The famous memory theorist, Endel Tulving, believed that material is more easily recalled if the context during retrieval is similar to the context during the initial coding of the material, a theory termed the *encoding specificity hypothesis*. This theory appears to have good empirical support. So students should study under the same or similar conditions under which they will have to recall material.

However, suppose you encoded something correctly and could recall the information at one time, but then forgot it. What explains that type of forgetting? One line of theories suggests that passive processes such as decay of the memory explain such instances. Other explanations are theories that tend to be more active, such as *interference inhibition* theories.

Forgetting as Interference

Proactive interference refers to difficulty learning new information because of interference from previously learned information, whereas *retroactive interference* occurs when someone has difficulty recalling previously learned information due to interference from newly learned information. The idea of a passive process that leads to the decay of memory traces has been hard to substantiate empirically; however, the active mechanisms have been supported in some of the literature. It appears that the ability to learn new information can sometimes be impeded by information already stored in the cortex, and recall of previously learned information can sometimes be interrupted by newly learned information.

> **Can a bump on the head restore memory?**
> While there are claims of this happening, this is most likely a myth, typically seen in films. Such an event has never been replicated experimentally to the author's knowledge.

Experiments looking at *directed forgetting* suggest that memories can be inhibited voluntarily. In some very clever experiments, participants are given two lists of words to remember. Some of the participants are later told that the first list was a practice list (after learning it) and it is okay to forget words from the first list, while others are told to remember both lists. Surprisingly, the recall of the list that the participants are instructed not to remember is significantly worse. These experiments demonstrate that the presence of strategic forgetting does occur. It may well be that once you learn material, use it to serve a purpose, and then decide either unconsciously or consciously that you no longer need to remember it, it is forgotten.

Learning Disabilities, Brain Damage, and Memory

Learning disabilities are categorized by domain. *Dyslexia* is a term that describes several different types of reading difficulties that are not related to intellectual, visual, or motor problems in the person. There are two basic types of dyslexia: an acquired form of dyslexia that is caused by brain damage to a person who was previously able to read well (the term for this is *alexia*), and developmental dyslexia, which surfaces when the child is learning to read (this is true *dyslexia*).

Developmental dyslexia is more common than acquired dyslexia, and the majority of research has focused on the causes of developmental dyslexia. As with nearly all disorders and all human traits, either behavioral or physical, there is a strong genetic component that contributes to acquiring dyslexia such that a person's probability of displaying some type of developmental disorder increases as a function of the presence of the same disorder in his direct relatives. Developmental dyslexia has a heritability of about 50 percent, which is not unusually high and is consistent with many other behavioral traits and developmental issues.

The difficulty in identifying the neural substrates in the brains of individuals with dyslexia is that there have been a large number of identified differences in the brains of these individuals, and no single brain-related pathology appears to occur in all or most of the cases. Many individuals with developmental dyslexia also have other attentional or sensory deficits linked to visual brain circuits or auditory circuits, or they have motor deficits. The most accepted explanation for dyslexia is that this disorder somehow results from a deficit in the ability to represent and comprehend speech sounds (deficits with *phonological processing*). The exact brain mechanism responsible for this deficit is not yet understood.

Other Learning Disorders

Dyslexia is not the only type of learning disorder. There are a number of different types of developmental learning disorders that have been identified, including such disorders as math-related learning disorders, nonverbal learning disorders, and others. Like dyslexia, there appears to be a strong genetic component to these, and like dyslexia, there are many hypotheses as to the neural substrates that are disrupted in these disorders; however, at this time the exact neural mechanisms involved in these disorders have not been verified.

Brain Damage and Memory

The term *amnesia* refers to a pathological loss of memory. Sometimes amnesia can be due to psychological stress or psychological trauma. Memory loss of some degree is a common effect of brain damage. This section will focus on brain damage as a result of traumatic brain injuries and surgical procedures. The progressive brain damage that occurs in certain neurological disorders, such as Alzheimer's disease, and brain damage due to conditions such as stroke will be discussed in a different section of this book.

Global amnesia refers to memory loss that is general in nature. The person is unable to remember information that occurred prior to her injury, during her injury, and after her injury. *Retrograde amnesia* refers to memory loss for events occurring before the brain injury happened. *Anterograde amnesia* refers to memory loss or an inability to remember events that occurred after the occurrence of the brain injury. The presence of lasting global amnesia or severe retrograde amnesia in individuals with a traumatic brain injury is actually rare, unless the brain injury is severe and the brain damage is extensive.

Most often following an injury to the brain, there is a period of *post-traumatic amnesia* where the individual is confused, may have global memory issues, and will often not remember the events that took place surrounding the time of the injury. As the individual recovers, it is not unusual for memories that were present prior to the brain injury to return, although there may be a *temporal gradient* associated with these memories such that information closer in time to the actual occurrence of the brain injury is more difficult to remember than information further back in time. It is not unusual for individuals who have even mild brain injuries to be unable to recall the event that led to the brain damage or events occurring within a short time before or after their injury. Anterograde amnesia or difficulties forming new memories is more common in cases of traumatic brain injury than is lasting retrograde amnesia.

Types of Brain Injuries

A *concussion* occurs when there is some type of mechanical force applied to the brain that results in any alteration of an individual's thinking processes. For most people, the typical bump on the head does not result in a concussion; however, many people do experience some form of mild concussion at least once in their lifetime. The experience of having one or two mild concussions has not been shown to result in any significant long-term effects (after all, nature built in recovery mechanisms in all animals for these events). Having multiple mild, moderate, or severe concussions in succession can result in long-lasting effects, and this is why there have been changes instituted in many contact sports, such as football, to protect the participants from this possibility.

Traumatic brain injury (TBI) is a condition that can occur when a severe mechanical force results in damage to the brain. Two general categories of TBI exist: closed-head injury, where there is no penetration into the skull, and open-head injury, where some projectile penetrates the skull and enters the brain. In the latter, such as gunshot wounds or other penetrating injuries, the effects of the injury are typically focal, that is, the behavioral effects as a result of the brain injury are typically limited to those associated with the area of the brain that has been penetrated. These effects may be temporary or longer-lasting, depending on the severity of the injury and other variables related to recovery in the person.

Closed-head injuries can have temporary or long-lasting effects and, depending on the initial severity of the injury, can have focal effects or more generalized effects. Because the circuits in the brain that deal with encoding, storage, and retrieval are so extensive, it is not unusual for individuals with moderate or severe TBI to experience long-lasting difficulties with memory. When the mechanical force applied to the brain is significant, this may result in not only injuries at the site of the blow to the head, but also in more diffuse injuries throughout the brain. When an object strikes the skull, it can produce damage to the brain at the site of the impact (called a *coup* injury), and the resulting shockwaves that travel through the brain and push it against the opposite side of the skull can result in injury in that area as well (termed a *countercoup* injury). Severe forces applied to the skull can result in rotational injuries, tearing axons in the brain, and diffuse brain damage known as *diffuse axonal injury*, which can lead to severe, long-lasting effects. In these cases, there may not be an actual physical blow to the head, but damage can result from forces that occur, such as during a car accident or when being violently shaken or tossed about (like in shaken baby syndrome).

> One issue regarding rehabilitation that does seem to have a good amount of empirical support is that having these patients exercise and remain active following their injury is related to quicker recovery times. Because no two people with TBIs have exactly the same injury, it is difficult to design empirical studies to compare the effectiveness of different cognitive rehabilitation methods.

The prognosis for individuals who have experienced some type of TBI varies depending on the severity of the TBI and on the length of any post-traumatic amnesia they may have incurred. Individuals with more severe TBIs and longer terms of post-traumatic amnesia typically are at risk for longer-lasting and more severe effects. In general, there is some recovery in most cases of TBI; however, many individuals display long-lasting or even permanent cognitive effects. There are certain medications and combinations of medications that have been demonstrated to aid in the recovery time of individuals who have cognitive deficits following a TBI. However, it appears that these medications are most effective when applied early in recovery and may not be as effective in the later stages of recovery.

The use of cognitive rehabilitation techniques in treating people with TBI is common, but there is still some question as to whether these methods offer any additional rehabilitative effects outside of normal recovery.

THE EMOTIONAL BRAIN

In his book, *The Emotional Brain,* Joseph LeDoux makes an interesting observation about emotions: "Unfortunately, one of the most significant things ever said about emotion may be that everyone knows what it is until they are asked to define it." Not only is the word "emotion" hard to define; it is also a difficult concept to measure, even though researchers are getting a better grasp of the biology of emotions, such as fear and anger. In this chapter, some of the more common theories about emotions, the brain, and behavior are explored.

Why Do You Have Emotions?

Most people think an emotion is an internal conscious condition. Most people also believe that while humans and even animals experience emotions, when it comes to computers or inanimate things, such as rocks and spoons, those objects do not have emotions. Consciousness seems to be essential before emotions can exist; however, no one can ever actually observe an internal experience in another person or animal. This limits the scientific study of many internal states, including the study of emotions.

One way to study emotions is to record observable behavior and compare it to or separate it from a person's reported feelings. This leads to operational definitions of emotions as behavioral responses (an operational definition specifies the operations used to produce something or to be able to measure it in research studies). Although emotions probably require some degree of consciousness, it is also possible to have an emotion without being conscious of what caused it. For instance, people with significant brain damage may develop emotional preferences to certain caregivers based on their experiences with them, but due to the brain damage, they may not remember the specific experiences that led to these feelings or even recall meeting the person previously.

The literal translation for the word *emotion* from Latin is "to stir up" or "to move." It is probably a safe bet that everyone has, at one time or another, advised someone to make a decision based on logic and not on emotion. It is certainly true that extreme emotional states can impair your ability to reason; however, approaching everything with a total lack of emotion, like Mr. Spock from *Star Trek*, is not helpful either. The most important decisions people make require them to make a prediction about what will make them feel good or bad, what is right or wrong, how their decision will affect others, etc. Individuals who experience certain kinds of damage to the prefrontal cortex often have a loss of, or blunted, emotions, and the quality of their decision making is markedly affected. In clinical cases of individuals suffering prefrontal cortex damage, those who—after their damage—outwardly express few emotions typically make shockingly poor decisions, often losing their jobs, ruining their marriages, and wiping out their financial savings (of course people with emotions make the same types of decisions, but prior to the brain damage these people had no such history).

Individuals who suffer severe prefrontal cortex damage early in life have also been shown to make poor lifelong decisions and to experience major hardships related to these decisions, even though they typically perform well on IQ tests. Often, such individuals express little remorse or guilt over taking advantage of others or over the many obviously poor decisions they make. So it appears that emotions have functions related to decision making. Emotions appear to guide important decision making strategies and can serve as important mediators of choice.

Emotions and Cognition

A long-standing controversy regarding the definition of emotion concerns whether emotions can exist without cognition. Many early theorists argued that in order for one to experience an emotion, one must also cognitively appraise his environment. In order to experience fear, a person must attribute the related properties of this emotion to a specific object or situation. However, not all theorists agree, and several important studies of emotion support the idea that emotions can be experienced without knowing what caused them.

> Critiques of research suggesting emotions can be identified from facial expressions suggest that a facial expression may not represent an emotion, but instead communicate a social signal. For instance, a smile could be interpreted as an expression of joy or an expression of a threat, depending on the context. Emotional states may also require a context.

Another assumption about emotions is that they may be expressed universally. Early research suggested that there were six types of emotions and that all other emotional reactions were variations or combinations of these six basic types (later theories identified more or less than these six core emotions). The six types are:

- Anger
- Happiness
- Fear
- Sadness
- Surprise
- Disgust

Research suggested that isolated people with no exposure to Western ideals can recognize some of these basic emotional categories from pictures of the facial expressions of Westerners.

Attack and Escape

Strong emotions are an incentive for action: When afraid, people try to escape; when angry, they want to fight. Emotions are connected to the stimulation of one of the branches of the autonomic nervous system, either the parasympathetic or sympathetic nervous system. These branches of the peripheral nervous system are tied to the "fight-or-flight" response. While each of these divisions of the autonomic nervous system contributes to both of these behaviors (fight-or-flight type readiness), in general, the parasympathetic nervous system is associated with slowing down bodily functions and the sympathetic nervous system with speeding them up.

Old and New Theories

One of the earliest and best-known theories of how the autonomic nervous system contributes to the experience of emotions is known as the *James-Lange theory*. In this theory it is the arousal from the autonomic nervous system and actions of the skeletal muscles that label one's emotions. A better way of putting this might be to say that people are afraid *because* they are running away, or they are angry *because* they are striking back. This theory is counterintuitive to what most people think happens.

In response to this discrepancy between the traditional understanding of emotion and the James-Lange theory, the famous physiologist Walter Cannon, the man who pioneered the understanding of the autonomic nervous system, argued that the sympathetic nervous system responds too slowly to be able to cause emotional states, and that both divisions of the sympathetic nervous system function as a unit. He believed that the sympathetic nervous system responded the same way to fear and anger and therefore could not precede the experience of emotion. The *Cannon-Bard theory* proposes that an event elicits both an emotional experience and physical arousal stimulation simultaneously, but independently. The following figure compares the three views:

> According to the James-Lange theory, the response of the body precedes the emotion and each emotion produces a different bodily response. A person can tell whether she is angry, afraid, happy, sad, etc., by noticing what her body is doing. This view goes against common-sense notions that the situation leads to an emotion, which then leads to the action.

Commonly Held View of Emotion:
Threatening Condition ➔ *Fear* ➔ *Physical Symptoms* ➔ *RUN!*

James-Lange Theory:
Threatening Condition ➔ *Physical Symptoms; Running Away* ➔ *FEAR!*

Cannon-Bard Theory:
Threatening Condition ⎰➔ *Fear*
⎱➔ *Physical Symptoms; Running Away*

More recent information has found that the sympathetic nervous system does not react equally in all emotional instances and that the brain receives information from multiple sources, such as hormones, skeletal muscles, etc. The autonomic nervous system responds differently to different emotionally charged situations. So the Cannon-Bard theory has been discounted.

According to the *Schachter-Singer Two-Factor Theory*, the physiological changes in your body inform you as to how strong a particular emotional state is (physiological response), but cognition, or your *cognitive appraisal*, of the situation informs you as to which emotion you are experiencing (cognitive response). In an interesting series of experiments, participants were given an injection of epinephrine (adrenaline) and told that they were receiving a new drug called "Suproxin" to test their eyesight. Epinephrine typically increases blood pressure, heart rate, and respiration, thus acting as a stimulant; however, not everyone knew what to expect when getting the injection. The researchers divided the participants into three groups and informed each group about the effects of the drug in different ways:

1. The *epinephrine informed group* was told about the effects and duration of the drug so that they would know what they would feel from the injection.

2. The *epinephrine ignorant group* was not informed about any potential symptoms of the drug.

3. The *epinephrine misinformed group* was given the wrong set of symptoms. They were informed that their feet would become numb, that they would experience an itching sensation over parts of the body, and that they would have a light headache.

4. Following being informed and injected, an actor came into the room and began acting either euphorically or angrily as the subjects were filling out a questionnaire. The actor made comments about the questions in either an angry or very happy way. Findings indicated that the participants who were in the misinformed or ignorant condition behaved similarly to the actor and reported feeling the same way the actor felt, whereas those in the informed condition demonstrated no emotional pattern. This suggests that participants who were informed about the drug attributed their feelings to the physiological effects of the epinephrine, whereas the uninformed or misinformed groups were not able to do this and interpreted the feelings as an emotion that was labeled based on the context. Schachter has replicated this finding under numerous conditions. However, sometimes emotions are experienced without knowing why, and other researchers have not always replicated Schachter's results.

> The importance of context in emotional experience can be seen outside the experimental arena. As discussed earlier, a smile can mean happiness or a threat for both the expresser and perceiver, depending on the context. Think of all the different contexts in which crying can express different emotions.

Where in the Brain Are Emotions?

Emotions are not fully experienced in the brain. Certain physical states contribute to the experience of emotion, and the perception of the context within which you experience these physical states contributes to the emotion you feel. It is important to understand that research in cognitive neuroscience has established several important points in understanding emotion:

1. Neuroimaging studies have determined there is not a specific brain center for each emotion. The brain activity associated with each human emotion is diffuse and not specific; therefore, patterns of brain activation appear to be associated with emotions.

2. There is almost always brain activity in the motor and sensory cortices whenever a person experiences an emotion or when a person is sympathetic to another person experiencing an emotion. This suggests that emotions are felt in the brain and in the body.

3. The patterns of brain activation that are recorded during the experience of an emotion are very similar to the patterns associated with imagining an emotion, or seeing someone else experience that emotion.

Structures of the Brain and Emotion

An area in the forebrain known as the *limbic system* is traditionally considered critical for the experience of emotions and may be important in overall consciousness as well. The limbic system consists of the amygdala, fornix, hippocampus, thalamus, hypothalamus, cingulate gyrus, and several smaller structures. These structures all border the brain stem and are also involved in other forms of processing, such as in memory and sensory processing in the case of the hippocampus and thalamus, and in fluid and temperature regulation in the case of the hypothalamus.

The amygdala is a collection of smaller structures in the anterior temporal lobe and appears to be associated with stimuli that have a particular emotional salience associated with them, especially stimuli that produce fear. The amygdala receives many inputs from subcortical structures, sends outputs to several subcortical structures like the hypothalamus, and sends outputs to the orbitofrontal cortex. The amygdala also appears to be involved in avoidance responses and the ability to associate specific contexts to threatening situations or even painful situations. The role of the amygdala has been studied extensively in classical conditioning, especially fear conditioning in animals, and is considered crucial in the acquisition and maintenance of fear-related behaviors. Some studies also suggest that the amygdala may be involved in the perception of certain positive stimuli.

The hippocampus has been discussed previously and is an important structure in mediating information from short-term or working memory into long-term memory. It receives inputs from nearly every area of the brain and sends numerous outputs back to numerous areas of the cortex. The fornix is a tract connecting the hippocampus to the hypothalamus and to other areas of the brain.

The anterior cingulate cortex is located on the frontal portion of the cingulate gyrus, which is just above the corpus callosum. It appears that this area is important in monitoring errors and conflicts during behaviors and in making corrections. This area also appears to be activated by the expectation of receiving a painful stimulus and is crucial in the neural mechanisms associated with reward and punishment, based on a particular behavior.

The orbitofrontal cortex is the anterior and medial part of the prefrontal cortex over the eyes. This area performs a number of monitoring tasks, especially tasks involving some form of risk or reward assessment (many researchers believe that this area of the brain is also involved in ethical decision making and in moral judgments). Damage to this area in humans often results in the person becoming abusive or impulsive, and is responsible for—but does not seem to interfere with—overall performance on IQ tests.

The experience of emotions and the interpretation of emotionally laden events or stimuli appear to be the result of these areas being stimulated and their communications with one another. These communications have extensive connections to one another and with other brain association areas that allow you to interpret the context of the events, and to then label and experience an emotional state associated with this context. In some situations this analysis of context may not be necessary, as in the classical conditioning of fear and in the reward-reinforcement-punishment circuits in the medial areas of the brain that act in an automatic fashion. The brain sends outputs to the muscles and glands, stimulating them to act, but there is also a feedback loop that sends information from the body and autonomic nervous system to the brain. These circuits often work in concert.

Stress and Emotions

A *stressor* refers to some type of experience that induces the *stress response*, a series of physiological changes in the body. Most people refer to the stress response as just plain old *stress*. While many people generally consider stress harmful or unnecessary, the fact is that you could not go through life without experiencing some form of stress. The continued exposure to chronic psychological stress has been associated with changes in physical and mental health; however, the experience of stress is a fact of life. Not all stressors are equal. The psychologist Richard Lazarus posited that the response to stress, positive or negative, determines its effect. There is a positive cognitive response to stress that is healthy and leaves a person feeling fulfilled, and there is a negative response that leaves one feeling drained.

General Adaptation Syndrome

Perhaps the most enduring of the theories regarding how chronic stress affects the body comes from the famous endocrinologist Hans Selye. Selye termed positive stress *eustress* and negative stress *distress* (referred to simply as *stress* in this chapter). Selye was interested in the effects of the bad stress and termed the process of reacting to stress the *General Adaptation Syndrome*. According to Selye's model, short-term exposure to stress produces adaptive changes in your physiology that allow you to respond to the stressor. However, if stress continues over the long-term, it produces maladaptive changes, and a breakdown of the system occurs. The General Adaptation Syndrome consists of three stages:

1. **Alarm.** Upon encountering a stressor, the organism reacts with the "fight-or-flight" response, leading to activation of the hypothalamic-pituitary-adrenal axis (HPA axis). In order to meet the threat of danger, hormones such as epinephrine and hydrocortisone are released into the bloodstream, mobilizing the body's resources to cope with the stressor.

2. **Resistance.** The nervous system acts to restore physiological functions to normal levels while the body focuses its resources to resist the stressor. There may be increased heart rate, respiration rate, and blood pressure as blood levels of glucose remain high and epinephrine and hydrocortisone continue to circulate through the system in elevated levels; however, the organism appears normal.

3. **Exhaustion.** If the stressor is chronic and continues, the organism's capacity to deal with it exhausts the body's resources, and the organism becomes susceptible to stress-related illnesses, disease, and even death.

For Selye, the body's response to the physical activation of the HPA axis was important in the response to stress; he largely ignored the contribution of the sympathetic nervous system in his original model. More recent models of the stress response suggest a two-system model, not unlike Selye's model, but expanded.

The major contribution of Selye's model is that both physical and psychological stressors affect the body in the same manner; however, the notion that there is only one response to stress, as proposed in Selye's model, is a major oversimplification. In modern models stressors can activate either the HPA axis or the sympathetic nervous system, resulting in large amounts of epinephrine and norepinephrine (adrenaline and noradrenaline) being released from the adrenal medulla in the adrenal gland just above the kidneys. The responses to stress can be very complex and variable, depending on individual differences in the person, the timing of the stressor, and the reaction of the person to the stressor.

One of the important hormonal contributions to the body's response to stress involves epinephrine (adrenaline), which is secreted from the adrenal glands. The adrenal glands also release other hormones in response to stressors, such as steroidal hormones and norepinephrine (noradrenaline). The response to stress involves the expenditure of energy. Epinephrine is involved in releasing glucose that is stored in the muscles (glucose is the body's most important source of energy); norepinephrine and the associated release of epinephrine help direct blood flow to the muscles, preparing the body for action (fight or flight).

Psychosomatic disorders are medical disorders believed to have psychological causes. Gastric ulcers were one of the first disorders classified as psychosomatic. Discovery of the bacteria H. pylori in most ulcer patients almost changed this thinking until the discovery that the bacterium alone could not cause ulcers. Gastric ulcers are more common in people with stressful lifestyles, and stress has produced gastric ulcers in laboratory animals.

Hormone Release

Hydrocortisone is a steroid secreted by the adrenal glands and is technically termed a *glucocorticoid* because it affects glucose metabolism. Glucocorticoids also break down protein to provide energy and convert fats into energy sources. Prolonged secretion of glucocorticoids has been shown to lead to high blood pressure, tissue damage, dysfunctions in the immune system, and slower rates of healing. More recent research has discovered that brief stressors result in physiological reactions that participate in the body's inflammatory responses. Brief stressors produce an increase in blood levels of a group of peptide hormones released by many cells called *cytokines*. Cytokines interact in a large number of immunological and physiological responses to stress and stimuli, and often result in a fever and inflammation of body tissues. Long-term stress can result in these effects being continued.

How Stress Affects the Immune System

One of the moderating factors of the effects of stress is how much control the person feels over the stressor. Even in animal models of stress, when the animals are allowed to control a stressor, the effects of the stressor are minimized. A long-standing model of stress involves first evaluating the environment for a threatening stimulus and then assessing the ability to cope with the stressor if it is labeled as threatening. Threats perceived as being under the person's control often do not result in significant physical or psychological repercussions.

Psychoneuroimmunology refers to the study of the interaction of behavior, stress, and the body's capacity to fight infection. The immune system is the body's defense system that protects against infections from viruses, other microbes, and parasites. The immune system is a collection of organs and white blood cells that develop in the thymus gland in the upper chest and in the bone marrow. The immune system constantly evolves to protect against threats such as microbes and other invaders, as these are continually evolving to develop ways of infiltrating the body. There are two ways in which the immune system responds to an invasion from these outsiders: The immune system recruits specific types of cells to fight an infection (cell-mediated immunity), or it produces specific chemicals to fight the infection (chemical-mediated immunity).

Glucocorticoids, released by the body when one is exposed to stress, can suppress the function of the immune system. All white blood cells contain glucocorticoid receptors. The secretion of glucocorticoids is controlled by the brain, and it appears that the hippocampus contains a large number of corticosteroid receptors. Brief stress activates the immune system, but long-term stress is associated with reduced immune system functioning. Many studies have reported that people experiencing long-term stressors are more prone to develop infection or disease. For example, medical students have been observed as being far more likely to contract an illness during examination periods, a very stressful time, than before their examination periods. In an interesting study, individuals were exposed to two common cold viruses and were monitored to see who would develop an infection that led to cold. Over 80 percent of the group became infected with the virus, but only half of them developed an actual cold. Those individuals who developed a cold from the virus were found to have chronic stressors lasting for at least one month, whereas those experiencing stressors of less than a month were significantly less likely to develop the common cold (other risk factors, including smoking, lack of exercise, poor sleeping habits, and poor nutrition, were also associated with getting sick).

Do Studies Show Causation?

It would be a mistake to think that the previous study "proves" that stress causes immune system breakdown leading to increased incidence of disease. Such a relationship is not that easy to demonstrate. For example, in the previous study there were a number of risk factors associated with an increased risk of getting a cold. People under severe stress change their routines: They may get less sleep; they may eat poorer; they may drink or use drugs, etc., and any one of these factors could be causal. Other factors can muddle the relationship between stress and the immune system, such as:

- The immune system has many redundant components, and so disrupting one component may not affect the others.

- No one knows how long changes to immune functions, produced by stress, last.

- A decline in one aspect of immune functions may be compensated for by increases in other aspects.

Because most of the studies involving immune functioning and stress are correlational (they show relationships but not causes), causal connections cannot be determined. Overall, there is evidence that chronic stress is associated with a greater propensity to develop health problems; however, the exact mechanism through which this works has not yet been identified. There are probably several different reasons for this relationship.

Stress and the Brain

Chronic stress appears to affect the brain in a variety of ways; however, the hippocampus appears to be particularly susceptible to the effects of stress. This susceptibility is possibly due to its dense number of glucocorticoid receptors, which appear to be involved in terminating the physiological responses to stress. However, chronic stress that leads to chronic elevation of glucocorticoid levels in the hippocampus results in a reduction of dendritic branching.

Exposure to stress early in life can have a variety of adverse effects. Children that are maltreated or subjected to severe forms of stress can display a variety of brain abnormalities. Some psychiatric disorders are believed to result from an interaction between severe stress and an inherent or genetic vulnerability. Being exposed to severe stressors at an early age can also lead to an increased stress response later in life. Studies with laboratory animals that expose pregnant females to severe stressors demonstrate delayed or disruptive embryonic development in the young.

> Studies of monkeys reared with wire surrogate mothers have indicated that contact, and not the provision of food, is what the young adhere to. In these studies the young monkeys clung to a warm comfortable surrogate and only visited a wire surrogate that provided food when they were hungry. Monkeys reared in isolation also developed severe neurotic problems as adults.

Other studies have found that physical contact with animals during the first weeks of life may produce resistance to stressful events. For instance, researchers found that when experimenters handled rat pups for a few minutes a day during the first few weeks of their lives, the rats as adults exhibited smaller increases in circulating glucocorticoids when stressed. Although it could be that the handled pups were groomed more by their mothers (as observed by the researchers) and this led to these effects. Other studies have shown that contact between the mother and infant results in a number of positive effects, whereas early separation from the mother appears to have negative effects.

Post-Traumatic Stress Disorder

One of the best-documented examples of the effects of stress on the brain comes from the psychiatric disorder known as post-traumatic stress disorder (PTSD). PTSD is a severe anxiety disorder that develops after a specific psychological trauma that involves having a severe threat to one's physical, psychological, or sexual integrity, or after witnessing such an event.

The stressor (which must be very severe in nature) overwhelms the person's ability to cope and results in a psychiatric disorder that is characterized by frequent episodes of reexperiencing the original event, dreams or nightmares regarding the event, isolation, attempts to avoid similar situations, hypervigilance (an enhanced state of sensory sensitivity and an exaggerated intensity of behaviors designed to detect threats), and other related symptoms. Interestingly, the preferred treatment for this disorder is a type of psychotherapy known as exposure or desensitization therapy, where the sufferer is guided through mentally reexperiencing the event in a relaxed state. Two of the most common stressors related to PTSD are wartime experiences and rape.

> Do not misinterpret these findings to mean that PTSD *causes* these changes; it is just as likely that these brain states are pre-trauma risk factors for developing PTSD, or that some other factor leads to both PTSD and altered brain functioning. In fact, recent twin studies have suggested that reduced hippocampal size may be a preexisting, and potentially predisposing, factor for PTSD.

Physiological Changes in PTSD

PTSD is distinguished by a specific disruption of homeostasis and the associated changes in the CNS. The behavioral presentation of the disorder is universal, indicating that this disorder is associated with some specific changes in the brain. Empirical investigations covering the neurochemistry and neuroanatomy of the human stress response have led to an important understanding of these disruptions. A number of the neurobiological changes associated with PTSD may be a result of the consequences of the trauma, but other changes appear to reflect preexisting vulnerability factors. The research has investigated the cognitive and affective processes involved in mediating PTSD. This research has revealed that abnormalities in memory, disruption of the normal extinction of fear, hypervigilance, and avoidance are associated with changes in the neurobiology of PTSD sufferers. The use of functional and structural brain imaging, as well as studies of hormone secretion, have been helpful in understanding the brain correlates of these symptoms.

Findings indicate that the *hypothalamic-pituitary-adrenal axis* (HPA axis), which is responsible for regulating stress hormone release, may be chronically activated in PTSD. Many structural imaging studies of the brain have reported decreased hippocampal volumes in PTSD patients, ranging from 5 to 26 percent decreases. The prefrontal cortex is involved in the inhibition and control of responses, attention, and other aspects of executive functions. Studies have indicated that the prefrontal cortex has been affected in people with PTSD such that there are prolonged states of anxiety and fear occurring in PTSD sufferers. There has also been evidence to suggest lower bilateral amygdala volumes are associated with the diagnosis of PTSD.

Stress-Increased Vulnerability and Diseases

It is important to note that studies have linked PTSD to an increased vulnerability to develop such conditions as cardiovascular disease, diabetes, gastrointestinal disease, fibromyalgia, chronic fatigue syndrome, musculoskeletal disorders, and other diseases. However, it is also important to note that a diagnosis of PTSD increases vulnerability to substance abuse, decreased personal hygiene habits, relationship issues, isolation, and other psychological disorders, such as depression. So it is unclear how these relationships affect one another; however, the association between extreme stress and vulnerability to health issues is certainly real.

DON'T BE SO SMART:

Intelligence and the Brain

HOW MUCH IS 2 × 4? Who wrote *Slaughterhouse-Five?* How do you tie a shoe? How are a pear and a chicken alike? *Intelligence* is said to separate humans from other animals (although more broadly defined, intelligences have been studied in plants and animals). There is no sound agreement on what actually constitutes intelligence; for example, are there many different types of intelligence or just one? This chapter covers some of the more popular theories of intelligence, how intelligence is expressed in the brain, and how consciousness relates to intelligence.

What Is Intelligence?

The conception of "intelligence" is actually a relatively new concept that was not defined until about a century ago. The word *intelligence* comes from the Latin word root *inter*, which means "between" or "within," and the Latin root *legere*, meaning "to pick out, bring together, choose, catch with one's eye," or "to read." Thus, *intellegere* can be interpreted as "to perceive," "see into," or "to understand." Most people define intelligence by referencing examples of behavior. When one references a behavior in relation to intelligence, one is conceiving intelligence as being adaptive in nature, a common conceptualization among intelligence theorists. One of the most important aspects of intelligence is that it allows for predicting future behaviors. An intelligent person can predict, with good accuracy, what the consequences of his behavior will be (of course, this does not mean the person will always *make* the best choice). Likewise, others can predict that an intelligent person will behave in a manner consistent with being able to adapt to the environment and understand his situation. Formal theories of intelligence have defined intelligence in a number of different ways, ranging from abstract thought to reasoning abilities to understanding emotions. Some of the more common conceptualizations of intelligence from major theorists are discussed here.

Galton and Cattell

English aristocrat and cousin to Charles Darwin, Francis Galton revived the term "intelligence" and made the first genuine attempt to measure intelligence. Galton was committed to the notion that there was a hereditary basis to intelligence, and he coined the popular term "nature versus nurture" to explain differences in innate abilities and learned ones. His intelligence-type tests were based on his notion that intelligence was a reflection of a person's neurological efficiency and thus inherited. Galton believed that intelligence could be measured through physical attributes, such as sensory perception, grip strength, and reaction time. In the late 1800s, Galton attempted to correlate such psycho-physiological measures with a person's rank in school or occupation. However, he found no such relationship.

Around the same time, James McKeen Cattell, a doctoral student from America, was performing reaction time experiments in William Wundt's laboratory in Germany. He became aware of Galton's work and began to communicate with him. Cattell developed his own series of psycho-physiological tests based on Galton's earlier ideas. The early efforts of both Galton and Cattell were very influential in experimental psychology and psychological testing.

Alfred Binet

Alfred Binet, a French psychologist, was challenged by public school officials to develop a test that would be able to identify children who were at risk of falling behind in their academic achievement. As a result, Binet, along with collaborator Theodore Simon, developed the first bona fide intelligence test in 1905. Binet also developed the forerunner to the notion of the "Intelligence Quotient" or IQ, which was the concept of "mental age" or "mental ability." However, his concept of how one's intelligence was to be derived, as well as his overall concept of intelligence, has been modified.

The original Simon-Binet test was made up of an assortment of items, with the goal of measuring knowledge and skills that an "average" French schoolchild of a specific age would possess. Items consisted of such things as verbal knowledge of food, naming objects, placing weights in order, and other verbal and tactile tests. The test items were graded in terms of their difficulty according to one's age. Thus, items that an average twelve-year-old could answer should be missed by a younger child of six. The difficulty level was determined by administering the tests to fifty normal children aged three to eleven years old and to some mentally challenged children and adults. By determining the average score on the test items for children of specific ages, Binet was able to determine what the average child of a certain age should be able to complete. He then could designate a specific child's "mental age" by her test performance on the same items compared to the performance of his standardization group (Binet did not like the term "mental age," instead preferring "mental level"). Children with mental level results significantly lower than their chronological age were identified as at risk of falling behind in school. No overall formulation or overall score was calculated as in current IQ tests.

Binet's approach to the understanding of intelligence was functional; he wanted to predict something. One flaw that is often noted is that his original tests included many verbal items and only some physiological measures. This reflected Binet's belief that intelligence was the capacity to be able to make judgments about relationships, comprehend meanings, and draw conclusions from relationships, all of which are primarily verbal in nature.

Charles Spearman

Charles Spearman was a student in Wundt's psychology laboratory. Spearman believed that intelligence was a general ability involving mainly the ability to see relations and correlates. Like both Cattell and Galton, Spearman preferred the notion of a biologically based, single source of human intelligence, even though earlier research had not found statistically significant relationships (correlations) between mental tests and physical capabilities. Spearman demonstrated that uncorrected correlation coefficients underestimated the actual strength of the relationship between any set of variables, and that such an underestimation is severe if the tests have a restricted range of values, like Cattell's tests had. (An intelligence test item score is restricted to a limited number of values.) Spearman developed a statistical correction formula and used it on Cattell's data, resulting in significant positive correlations among the mental tests and physical attributes. Using his factor-analytic method, he was able to demonstrate that all intelligence test items loaded heavily on a single factor, which he named the "general factor" or as it is known commonly, g. This factor still forms the foundation for many current theories of intelligence.

Lewis Terman

During World War I there was a need for the U.S. Army to develop a quick-administration intelligence test for deciding the type of training that a recruit would receive. Psychologists Robert Yerkes, Lewis Terman, and others worked together and developed the Army Alpha and Army Beta tests. The Alpha test stressed verbal abilities, whereas the Beta test stressed nonverbal abilities. The Beta test was administered to those performing poorly on the Alpha test. After the war, these tests became the models for future intelligence tests. After World War II, Terman, who was at Stanford University, translated the Binet-Simon tests into English and adapted them for American schools, renaming the test the Stanford-Binet Test. Terman later took the notion of mental age and one's chronological age to compute a new metric called the IQ via a ratio method: (Mental Age/Chronological Age) × 100 = IQ. This new score was consistent with Spearman's notion of g and Terman's view of intelligence as the capacity to form concepts and to grasp their significance. The current derivation of an IQ score is not based on the concept of mental age, but is calculated by comparing how one scores in comparison to how one's *reference group* performs on the test (a reference group is comprised of people with the same demographic characteristics).

David Wechsler

Psychologist David Wechsler did not believe that available intelligence tests were accurate measures as they overemphasized verbal abilities (e.g., Binet's tests). Wechsler devised probably the most widely used series of intelligence tests that are in some ways similar to the Stanford-Binet test. However, he also included a number of nonverbal or performance tasks based on the Army Beta tests. The original Wechsler tests provided three separate IQ scores: Verbal IQ, Performance IQ (based on nonverbal items), and a Full Scale or overall IQ based on the other two scores. The Full Scale IQ score was similar to Spearman's g. Wechsler also used a different form of standardization that was based on the normal distribution and allows for the expression of IQ in terms of standard deviations from the mean or average IQ score. This allows the same IQ score of different age groups to have the same percentile rank, which was not the case in the other methods of calculating IQ scores, such as in Terman's ratio method.

Howard Gardner

Some theorists have objected to the notion that intelligence represents a unitary construct. Perhaps the most well-known of these is Howard Gardner's theory of multiple intelligences. Gardner believes that intelligence tests are too restrictive and has outlined at least ten different intelligences, ranging from spatial and linguistic to moral intelligences.

There have been many criticisms of Gardner's approach. First, Gardner provides no empirical support for his designations, and there is none elsewhere to support them. Second, psychometric studies have constantly supported the notion of a unitary factor common to intelligence tests that is made up of many related functions. Third, modern intelligence tests measure many different abilities that reflect an overall underlying factor. What Gardner has done is to substitute "intelligence" for certain abilities and to try to demonstrate that these abilities are mutually exclusive forms of intelligence, when in fact, they most likely reflect different aspects of an underlying general construct.

> Gardner purports nine separate intelligences: spatial, linguistic, logical-mathematical, bodily-kinesthetic (such as athletics), musical, interpersonal (social abilities), intrapersonal (self-knowledge), naturalistic (relating information to one's natural surroundings), and existential (the ability to consider information transcending sensory inputs, such as infinity).

Cattell-Horn-Carroll Theory

The Cattell-Horn-Carroll theory of cognitive abilities is a hierarchical model of intelligence that has received solid empirical support. The original theory proposed two broad categories of intelligence: *crystallized intelligence*, which involves knowledge that comes from prior learning and past experiences, and *fluid intelligence*, which is the ability to perceive relationships independent of previous practice. Carroll expanded on the Cattell-Horn theory and proposed a model that contains over seventy specific abilities that can be categorized under eight primary second-order abilities. The eight primary abilities are all part of an overall *g* ability (general intelligence).

Are There Different Types of Intelligence?

Suppose that Jerry is a baseball player. Jerry can hit home runs, but he has a low batting average (for baseball people, suppose Jerry hits .250). In addition, Jerry can throw the ball hard, but his overall fielding skills are below average. One might say that Jerry can hit home runs, but he is an average hitter. Because Jerry's overall baseball skills are not well developed, it would be wrong to say that Jerry is a good baseball player. Hitting is an aspect of the game of baseball, as is fielding. They are skills that comprise the concept of "baseball." Likewise, intelligence is conceived of as an overall concept that is composed of many different skills. A person may be a good mechanic, but he may also score poorly on intelligence tests. Like Jerry, he may have a specific skill that he is very good at, but he may not be conceived of as an "intelligent person."

One of the observations that should become apparent from discussion of the general theories of intelligence is that most of them refer to an overall or general intelligence factor (often called g as Charles Spearman referred to it) that is comprised of other specific abilities. This does not necessarily mean that there are different *types of intelligences*, as Howard Gardner labels them, but that the concept of intelligence is composed of many different types of abilities that are separate and yet related. Because there is much variation in behavior, most people will exhibit strengths in some abilities and relative weaknesses in others.

> Decision making depends on a number of factors, and while it is aided by intelligence, it is not a direct product of it. Some very creative people have performed poorly on intelligence tests, and some very intelligent people have made some very questionable decisions. In general, intelligence is one of the best predictors of life outcomes, but it is not a perfect predictor.

A distinction made by laypeople is one of being "book smart" or having "common sense." This certainly can be a distinction; however, most intelligence tests have a component that measures common-sense reasoning. Furthermore, research has indicated that people who score higher on intelligence tests generally do better in life, such as having better jobs, being more satisfied with their lives, and being better adjusted. The important thing to understand about this relationship is that it is not perfect; not everyone who scores well on intelligence tests does well in life or makes good decisions, but many more of them do.

Components of Intelligence

One of the reasons to understand what intelligence is, and what the components of intelligence are, is to allow people to develop and use their intelligence constructively. When the concept of intelligence is better understood, researchers can identify what aspects of intelligence can be modified.

One interesting relationship regarding intelligence across all species of animals is the relationship between brain size, body size, and intelligence. The larger the *ratio* of brain size to body size, the more intelligent the species tends to be, although this is not always the case. There also seems to be a link between intelligence and the relative size of the prefrontal cortex in animals. Primates and dolphins have the largest prefrontal cortices and also are the most intelligent of animals. If this is the case, then it follows that there is a relationship between genetics and intelligence.

As is the case with almost every trait—even behavioral traits—intelligence, as measured by IQ scores, appears to have a significant genetic component. The estimates of the heritability of scores on IQ tests typically are around the 0.5 range, although some studies report stronger heritabilities and others weaker heritabilities. Heritability refers to the proportion of a trait that is expressed by a particular genotype (set of genes) in a particular environment, but it does not refer to how much of a trait is *caused* by the genotype or genes. There are two important things to take home from this finding: First, all traits most likely have some inherent factor that contributes to their expression; and second, all traits are expressed via the interaction of genes and the environment.

Still, the relationship between genes and intelligence is real and undeniable. A person's genotype does determine in part how well that person will score on an IQ test. Probably the best way to view a genetic component to intelligence is that it helps to define the capacity of the person to express that particular trait. Genetic contributions define the upper limits of the trait (e.g., intelligence) that the individual can reach given a "perfect" environment. Moreover, very small differences in genotypes can be enhanced by experience and upbringing. Therefore, environment is the modifiable component of intelligence. Also, people who are inherently more gifted at learning will often seek to learn, thus furthering their mental abilities, whereas someone who inherently has trouble reading, for example, may not read as often. Thus, the genetic and environmental components of intelligence are interrelated and basically inseparable.

IQ and Intelligence

IQ tests were originally designed to measure a student's potential to do well in the French public school system and then were applied in other countries. Over the years, and after many modifications, they have come to be identified with intelligence. A few years ago, a controversial book purported that certain ethnic groups demonstrated overall lower performances on IQ tests than other groups and that intelligence was inherited. As previously discussed, intelligence depends on experience and inherent factors. Of course when educational and socioeconomic backgrounds and opportunities are equalized between different groups, many of these aforementioned differences in IQ scores disappeared or were significantly reduced.

> Studies of rodents in enriched environments have demonstrated how the brain is affected by the stimulation provided by its surroundings. Rodent brains from animals exposed to more enriched environments have demonstrated increased numbers of synapses and thicker cortices. This effect has been documented in rats placed in these environments that were the equivalent of eighty, and even ninety, years old!

Constructs

The first thing to understand is that while IQ is a measure of intelligence, IQ is not the same thing as intelligence. Intelligence is a complex, abstract, psychological construct. Constructs are explanatory concepts that are not directly observable, such as things like thought, personality, feelings, the number three, pi, and intelligence. What are observed are the results of these "things"; how they affect the world, and not the construct itself (for instance, you can show someone three apples, three clocks, or three brains, but you can never show just *three*). Constructs are used to develop scientific theories, describe events, make predictions, and classify things. IQ tests are used to predict a person's ability to function in environments such as school, or to determine if a person may have some deficit in functioning related to brain damage or a developmental issue. Intelligence, which is in part measured by IQ tests, is a broader construct.

People who are at the significantly low end of the IQ spectrum (IQ scores of 65 and below) were in the past designated "mentally retarded." That term has lost favor, and the appropriate term is now having a "developmental disability." As a group, these individuals often need some form of supervision, depending on the level of disability. At the other end of the spectrum, intellectually gifted individuals often do well in school and life, but not always, as IQ does not predict things like motivation or measure things like self-control or emotional stability. People who fall between these extremes are capable of quite a bit of variation in performance.

Mental retardation (MR) is defined by an IQ score of 65 or lower and significant developmental impairment. IQ levels for MR are: mild (50–69, can be educated to the level of an eight- to eleven-year-old); moderate (35–49, trainable to attain a four- to seven-year-old level); severe (20–34, requires lifelong supervision, but can learn some personal maintenance skills); and profound (below 20, requires custodial care).

There are some individuals with overall low IQ scores who are known as *savants*. This is not a common occurrence in individuals with severe developmental disabilities, such as autism, but when this phenomenon does occur, such people typically have an extraordinarily well-developed mental capacity that is often limited to one domain. For example, the person may be able to calculate the day of the week or the specific date for any particular historic event or display extremely well-developed arithmetic abilities that rival the fastest computers. There seems to be no reliable neuroanatomical feature in these cases, and even in cases where the savant is not autistic, these skills seem to be counterbalanced by poorly developed skills in other areas that are necessary for normal living. Such skills do not fit the conceptualization of intelligence, but instead appear to be isolated special abilities.

How Are Consciousness and Intelligence Related?

Philosophers and psychologists have debated the meaning of consciousness for centuries. At one point it was believed that consciousness could not be defined; however, with the advent of sophisticated neural imaging techniques and other new technologies, the understanding of what it means to be conscious has broadened greatly. Probably the most commonly accepted notion of consciousness divides the notions of *pure consciousness* and *pure awareness*.

Awareness includes a certain state of perception that results in behavior, and that can be experienced. The difference between pure awareness and consciousness is mostly due to language. Consciousness allows for one to use language to relate one's experience and classify perceptions that are created by awareness. Consciousness is linked to memory such that, through language, one can link current experiences with prior concepts. Interestingly, a great deal of the research on intelligence has also associated intelligence with language abilities, despite the fact that different skills and abilities that contribute to intelligence are labeled as not being language-based.

The expression of intelligence in the brain is largely thought to be a function of the frontal and prefrontal cortex; however, the notion of intelligence requires the ability to use the entire brain, especially language-based abilities, which reside in the temporal lobes. Moreover, many severe injuries to the frontal lobes, such as a lobotomy, may not significantly affect IQ scores.

Recall that the human brain is composed of two hemispheres connected by commissures, the largest of which is the corpus callosum. Humans who have had the majority of their left hemisphere destroyed as adolescents or adults (recall that the left hemisphere is the dominant hemisphere for language) are often reputed to be profoundly mentally retarded and appear to respond to stimuli in an approach-avoidance manner. People with right hemisphere damage will often have poor visuospatial processing, but often appear to have relatively normal cognitive functioning. The famous neuroscientist Michael Gazzaniga developed the notion that the left side of the brain tries to make sense of the world by using language and contains an interpreter that is constantly making up a verbal story about experiences. Without the interpreter, awareness would exist, but it would be awareness without context, and behavior would be expressed in terms of stimulus-response associations. Thus, consciousness and intelligence are intimately related.

How Does It Come Together?

A number of neurobiological theories have attempted to relate conscious experience with neural activity, and there are many proposed neural correlates to conscious states. For instance, Crick's theory proposes that consciousness results from the activity of neural assemblies, which are collections of neurons. Crick's theory is represented by measuring the electrical activity of the brain and oscillations in the cortex, which form the basis of consciousness and are interconnected with different forms of sensory awareness. Binding theory refers to the process where separate pieces of information, about a single stimulus, are brought together and used for further processing later.

Unfortunately, Crick's theory has never been able to explain the importance of the brain oscillations that allegedly give rise to conscious experience, and the binding theory has never been empirically demonstrated. Where do all of these isolated experiences come together in the brain? There is no good answer. One hypothesis proposes that consciousness requires the involvement of the sensory cortical areas that include the frontal lobe, parietal lobe, occipital lobe, and the thalamus. A coma refers to a state of unconsciousness that is not sleep and typically lasts more than six hours. A coma typically results from severe brain trauma or another insult to the brain. A vegetative state occurs when a person comes out of a coma (his eyes open) but he otherwise does not appear to be responsive to stimuli. Studies using brain imaging methods have been able to determine that the sensory areas of the brain in comatose and anesthetized patients respond to familiar music via input relayed from the thalamus to the frontal lobes. However, in truly vegetative patients, there is no feedback signal from the frontal lobes back to the sensory areas of the cortex, as occurs in conscious individuals. These findings seem to support this hypothesis of consciousness. Damage to other areas of the brain, such as to the temporal lobes and frontal lobes, can also disrupt consciousness as seen in post-traumatic amnesia and other types of amnesia.

Unconscious Processing

The notion of an "unconscious mind" is the foundation of many psychological theories of behavior, especially those of Freud and his followers. However, Freud's ideas have been replaced by a different notion of unconscious processing, such as the large component of memory that is implicit in nature, in that the person often cannot consciously recall where the memory came from. One cannot usually recall when and how one learned to tie one's shoes, but one still can perform the procedure effortlessly. However, there is also empirical evidence to suggest that a fair amount of unconscious processing goes on in the brain directed at expressing attitudes and motivations. This has become referred to as *implicit cognition*.

> Priming is an implicit effect where being exposed to a stimulus influences a later response. For instance, reading or seeing "brain" and later being asked to complete a word starting with **br** results in an increased probability that one will say "brain" rather than "bring" or "broke." Priming is used to measure attitudes, understand brain processing, and judge the effects of brain damage.

Implicit attitudes are the positive or negative feelings, thoughts, or even actions that are directed at objects in the environment and that have developed due to past experiences. A person is either unaware of this attitude or cannot attribute it to a specific previous experience. These attitudes are often revealed in studies using specific measuring techniques of reaction time or choice preferences. For example, in a version of a test known as the Implicit Association Test, participants in an experiment make split-second decisions about whether or not to shoot intruders in a video game format. The decision to shoot or not to shoot is based on whether the video intruder is holding a gun or not. Many people exhibit an implicit bias toward blacks as measured by faster and more frequent decisions to shoot the intruder when he is depicted as a black man. Implicit cognition is studied in a number of contexts and relates to the automatic-processing mental component of the dual process model of cognition. It is fast and acts without conscious awareness.

YOU'VE GOT PERSONALITY

WHAT IS PERSONALITY? There have been many different conceptions of personality, both socially and biologically based. Because your personality is reflected in your behavior, it must have a biological basis in the brain; however, what you may not know is that the whole idea of personality theory was almost scrapped by leading psychologists at one time. Can you imagine not having any personality at all? (You may claim to know such people.) This section looks at what personality is, some various ways it is measured, and some theories about how it is expressed in the brain.

What Is Personality?

The word "personality" comes from the Latin word *persona* meaning "mask." Different personality theorists offer different variations defining what personality really is, but basically, personality is a set of organized and dynamic characteristics that influences your thinking, emotions, motivations, attitudes, interpersonal relationships, and behavior in a unique manner. Yes, that is a mouthful. However, a crucial aspect of the definition of personality is that it is *consistent*. In addition, personality must *do something*; personality must have a function. Your personality should express needs, define motivations and drives, allow for relationship building, etc.

> Modern trait theory is based on lexical descriptions of personality (traits taken from dictionary descriptions or others descriptions of personality) and has been subject to a statistical technique called *factor analysis* that groups related traits together. One of the current popular models in personality theory is the five-factor model that identifies five basic traits of personality.

A major assumption of any personality theory is that the expression of personality has to have consistency in order for it to be a valid concept. Personality theories of all types were quite popular even before Freud, and many of these theories were either trait or type theories. Traits are subcomponents of a personality that are relatively enduring (consistent) ways of perceiving and thinking about your environment over a variety of different situations, whereas type theories stress personalities as more or less a complete package. Traits like honesty, agreeableness, etc., can come in any number of combinations and degree, whereas types are general overall descriptions that are relatively invariant, such as being extroverted. Freudian notions of personality were based on drives and motivations that were unconscious, but Freud's idea of personality expression combined the trait/type dichotomy. His most famous follower, Carl Jung, leaned toward types. Modern interpretations of trait theory, using statistical techniques, have identified basic traits that have withstood empirical scrutiny.

The five-factor model or the "Big Five" model of personality traits has emerged as a very enduring model for understanding the relationship between personality and other behaviors (it predicts something). The Big Five factors are:

1. Openness (inventive/curious versus consistent/cautious)

2. Conscientiousness (efficient/organized versus easy-going/careless)

3. Extraversion (outgoing/energetic versus solitary/reserved)

4. Agreeableness (friendly/compassionate versus cold/unkind)

5. Neuroticism (sensitive/nervous versus secure/confident)

Each major trait is composed of numerous sub-traits.

You Almost Had No Personality

Personality theorists experienced big trouble when behaviorism became the most popular paradigm in psychological thought. Behaviorism eschews notions of enduring traits or types and focuses on environmental contingencies that shape behavior. According to the behaviorists, personality is not consistent, but behavior is specific to the situation; therefore, the person is reacting to the environmental contingencies and does not possess a "personality." Some situations encourage extroversion, some introversion. The behaviorists added that the term "personality" is just a description that people apply to a label person's actions, but it really does not exist. Believe it or not this notion almost led to the death of personality theory in psychological thought. When research was performed on individuals who scored high on trait measures like extroversion or introversion, researchers discovered that the individuals did not always act in such a manner, depending on the specific situations they were exposed to. In fact, the best predictor of how a person would act in any given situation was—the situation! This was exactly what the behaviorists would predict, and personality theory almost became extinct. Personality theorists spent years trying to counter these findings but could not.

Eventually the situation was resolved when personality theorists measured behavior over many different situations and averaged out the general tendency of a person's responses. What was found was that, in any single situation, a person's score on a personality test was not a good predictor of how she would act; however, when behavior was summed up across many situations, it was found that people generally acted in accordance with their scores on personality measures. Personality theory survived.

Personality traits or types are typically measured in theoretical research studies, not by self-descriptions, but rather by other people's descriptions of what behaviors mean. In clinical personality tests this situation is often reversed; the person's personality type is determined by self-descriptions. Both methods have obvious drawbacks.

Other popular theories of personality include social cognitive theories, which consider personality to be based on cognitions, such as perceiving the world and making judgments, or humanistic theories, which consider personality to be based on one's drive toward self-actualization. Most theories of personality incorporate traits and types, except for strict behaviorism. Biological theories of personality also incorporate traits and types but purport that these are biologically based as opposed to psychologically or socially based.

How Personality Is Measured

Two basic categories of personality tests exist. Projective tests present the person with an ambiguous stimulus or situation and require him to make sense of it. The Rorschach, or inkblot, test is the most famous of these. In this test, the person looks at an inkblot and relates what he sees. Responses are considered to be unconscious reflections of needs and desires. The test is evaluated based on how other individuals with defined traits have responded to these ambiguous tests. There are a number of different types of projective tests; however, the use of projective testing has generated significant controversy due to issues with poor reliability and with the validity of the interpretations made from these tests.

Issues in test theory include *reliability* and *validity*. Reliability is the consistency of the test results over time, over different people, or over different raters. It is expensive and requires expertise to develop a reliable personality test. Validity refers to the truth of the results: do the results actually measure what they say they do? It takes even more expertise to develop a valid personality test.

The other major category of personality tests is an objective test method, where an individual answers a question by choosing one of several choices (similar to a multiple-choice test). There are many types of objective personality tests, ranging from tests for clinical disorders, to tests for vocational aptitude, to frank measures of traits or types.

Objective tests are still controversial in some circles, but overall, they have better statistical properties and empirical support than do projective tests. However, not all objective tests are sound. The market is flooded with a number of tests that can be bought at the local bookstore or online, and many of these have little empirical evidence to support their outcome. Many of these commercially available personality tests operate on the P. T. Barnum effect, by making general conclusions that are applicable to most people (e.g., "You are considerate and caring, but sometimes get angry").

Phineas Gage: The Classic Study

Phineas Gage was a railroad crew foreman in the 1800s. In order to blast away parts of the landscape to make way for the railroad, workers typically ignited explosives with a long metal rod. On one occasion, Gage ignited some explosives, but the rod he was holding was propelled through his jaw and out through the top of his head, severing his frontal lobes from the rest of the brain. Gage lived and appeared to be none the worse for wear; however, his personality made an abrupt change. The rather mild, easy-going, reliable Gage became quite impulsive, foul-mouthed, and very unreliable. The case of Phineas Gage is an important landmark in acknowledging the idea that personality is fundamentally a function of the brain.

Biological Notions of Personality

Gage's case notwithstanding, early investigations of *temperament*, defined as natural dispositions of infants that endure into adulthood, have given credence to the notion of biological contributions to personality. Research in the 1950s, by Alexander Thomas as well as Jerome Kagan, has indicated that certain personality dispositions can be identified at a very young age in children and tend to endure over time. These types of studies rated the responses of infants and young children to novel situations. Thomas and his associates initially recognized nine dimensions of temperament that later were found to cluster in three groups: easy, difficult, and slow to adjust. Nearly 65 percent of the children they studied fit one of these categories. Other researchers have identified similar clusters of temperament. Temperament has often been cited as support for biological notions of personality in that there does seem to be a relationship between an infant's temperament and later personality; however, the development of the brain begins within the womb and continues well past birth. The so-called biological foundations of temperament are mediated by experience and interaction with significant others.

The *self-referential effect* refers to the observation that people remember events or information related to themselves better than they remember events or information related to others. This may lead to a discrepancy between how one views himself compared to how others perceive him. One may readily recall times when he was daring, but overall, he may be quite introverted. This may lead to discrepancies when subjective and objective methods are used to describe a person's traits.

Nature Versus Nurture and Personality

Information from EEG recordings has suggested that infants and children with different temperaments differ in amygdala and limbic cortex activations during unfamiliar situations. Difficult and slow to adjust children may have more excitable reactions in these brain areas leading to more anxious physiological states in novel situations. However, there is no way to control for experience in the studies, as exposure to novel situations begins at conception.

The personality traits of neuroticism (emotional instability) and introversion/ extroversion have been demonstrated to be consistent across cultures. Other traits have also demonstrated significant consistency over different cultures. If major personality traits demonstrate stability over different cultures, this suggests that much of what is described as personality may have an innate basis.

Cross-cultural studies of the five-factor model of personality also suggest that there may be a biological foundation for personality traits. Cross-cultural studies investigate how different cultures describe personality variables in people. If the results of these studies indicate similar trait descriptions occurring in totally different cultures, researchers infer that cultural differences do not contribute significantly to personality descriptions.

One interesting observation about temperament is that many theorists find evidence for four categories of temperament, which coincidently relates back to early Greek thought about personality being related to the four humors (fluids) in the body. The classic temperaments are: sanguine (blood), choleric (yellow bile), melancholic (black bile), and phlegmatic (phlegm).

Some researchers emphasized temperament as a method to understand how to educate children in the elementary school system. Many believe that temperament is most influential during these formable years, and that temperament diminishes in its effects on personality as the child matures. As people move into adulthood, they are capable of altering their temperament. Likewise, most theorists approach the nature versus nurture issue with personality as one of an interaction and not as dynamically opposed poles.

The interaction between biological and environmental influences can be understood in terms of mutual influences. An infant's temperament can affect how the parents respond to the child; this response can affect the child's temperament. For instance, a difficult child can elicit either more or less attentiveness by the parents; these parental behaviors can affect the temperament of the child, which in turn affects the parent's behavior, etc.

Personality Disorders

Personality disorders are defined as relatively enduring, rigid, and dysfunctional patterns of behavior that have been present since childhood or adolescence and continue on into adulthood. The current classification for these disorders indicates there are eight identified personality disorders in three related categories. While no biological basis for personality disorders has been determined, certainly research has indicated that people with certain personality disorders are associated with having different patterns of processing information than are people without personality disorders, suggesting biological differences. These differences in processing can be compared to brain volume differences and may give hints regarding how normal personality is expressed in the brain.

For example, antisocial personality disorder is characterized by a lack of empathy, the tendency to break rules and regulations, self-centeredness, and long-standing tendencies to use other people for one's own selfish purposes. Habitual criminal offenders are often antisocial types. Early studies on antisocial behavior indicated that many of these individuals processed verbal information differently and that verbal IQ scores in many of these individuals were lower than nonverbal IQ scores (the pattern is typically reversed in most people). Other research has suggested that the amygdala in people diagnosed with antisocial personality disorder is smaller than normal, and that this may contribute to their lack of empathy, issues with authority, and their tendency to engage in criminal behaviors.

In contrast, dependent personality disorder is characterized by feelings of nervousness, a need to be taken care of, the need for constant reassurance, and a marked inability to make decisions on one's own. As might be expected, parenting styles have been hypothesized to contribute to this disorder. Authoritarian parental styles may interfere with a child's experience of learning via trial and error, which is an important way young children learn. Authoritarian parents (strict disciplinarians with rigid rule systems) do not allow children to be autonomous. Overprotective parents can also lead to high levels of dependency in children in a similar manner. Overprotective parents can foster the belief that the child is unable to make decisions or cope without another person. Biological contributions include temperament such that infants with fearful, sad, or withdrawing type temperaments, and children with chronic health issues, may foster overprotective behaviors in parents, which may lead to dependent personality disorder. These findings highlight the interaction between biological factors (e.g., genetic makeup) and the environment in the development and presentation of one's personality.

Other neuroimaging studies find differences in the prefrontal cortices and limbic system in people with personality disorders compared to normal controls. But there is no identified brain scan prototype that can diagnose any personality disorder or any trait.

Personality disorders, like all psychiatric disorders, represent extreme departures from normal behaviors. A person with a personality disorder typically has a very difficult and often guarded prognosis, as these disorders are by definition lifelong disorders. The identification and diagnosis of a personality disorder or any psychiatric disorder should be done by a psychologist or psychiatrist and cannot be determined by untrained individuals.

Can You Rewire Your Brain or Change Your Personality?

So where in the brain is personality? In a recent study, researchers correlated brain volumes with participants' scores on the five-factor model of personality. The results indicated a number of predicted and unpredicted associations between the size of certain brain structures and the participant's score on a trait. What research like this indicates is that no one has a good idea of where personality traits are represented in the brain and that there are probably a number of areas that contribute to any one particular personality trait. Returning back to the case of Phineas Gage and the earlier discussion of how frontal lobotomy changed the personalities of people who had them, one can hypothesize that temperament and other biological foundations of personality are strong functions of the frontal lobes. Because the frontal lobes of the brain, the limbic cortex, and memory systems continue to develop, it can also be hypothesized that temperament and personality can change significantly as a result of a child's experience. One's personality would also be influenced by all sensory experiences and all interactions with others. Thus, personality would be a whole-brain phenomenon and be a result of a complex combination of innate factors and environmental interactions.

A theory that is applied to explain the cause of many psychiatric disorders is called the *diathesis-stress model*. This model purports that certain people are born with an inherent or genetic susceptibility (a diathesis), which by itself is incapable of causing a psychiatric disorder. The central idea of this model is that these susceptible individuals are exposed to stressful conditions at some time during their life, and the combination of the inherent susceptibility and stress produces a specific psychiatric disorder. The problem with this theory is that the specific inherent susceptibilities are difficult to determine by empirical methods and the definition of what actually comprises a severe stressor is also hard to determine. For instance, many people experience abuse as children, but not every abused child develops a psychiatric problem, and many different psychiatric disorders are known to be associated with moderate rates of being a victim of child abuse. In spite of extensive research, very few specific stressors have been shown to be related to the development of a specific psychiatric disorder, such as PTSD. A diathesis-stress model could also explain the development of many traits or personality types that are not considered disordered.

The idea that the brain is plastic indicates that brain rewiring is an essential function of the CNS. Brain rewiring occurs constantly in perception and learning. Whenever you learn something, you rewire your brain. However, it is well known in clinical circles that personality disorders are among the most difficult of psychiatric disorders to treat.

It follows that normal traits that people have are probably also difficult to change or to "rewire." In order for a full-blown change in a trait to occur, often one must experience a life-changing event or be highly motivated, as personality expression is often part of automatic processes and not something people always stop and think over. Personality change requires hard work. But people do change.

———◆———

THE ADDICTED BRAIN

THE TITLE OF THIS CHAPTER is a bit misleading, as it actually is a mistake to suggest that the brain becomes addicted. It is more accurate to say that the *person* becomes addicted. Addiction is a very serious problem worldwide. For example, it is estimated that in the United States, over 60 million people are addicted to nicotine, alcohol, or both. Many millions more people are addicted to other substances and even to certain activities, as the current conceptualization of addiction is not limited to substances and drugs alone.

Addiction (Dependence)

Most people do not understand what really qualifies as an "addiction." An addict may be someone who habitually uses drugs (or habitually engages in some other behavior); however, not all habitual drug users are addicts. The clinical differentiation between habitual users and addicts occurs when they continue to use the drug despite adverse affects on their social functioning, occupational functioning, or physical health, and repeatedly fail when they try to stop using the drug or engaging in the behavior.

The current diagnostic criteria for addiction, or as it is now termed "substance dependence," does not necessarily include physical symptoms, such as withdrawal, tolerance, and physical dependence (although these certainly are signs of addiction). In fact, if addiction were purely a physical phenomenon, then hospitalizing an addict for a short period of time and waiting for withdrawal symptoms to stop would result in an effective treatment. However, the relapse rate for addicts who leave detoxification centers without other supports and treatments is extremely high (relapse rates are high even with treatment). Therefore, addiction or dependence is a complex behavior involving brain systems, motivation, and elements of volition and choice.

Tolerance and Withdrawal

Tolerance refers to the notion that as a person uses a drug repeatedly, or engages in an activity repeatedly, it takes more and more of the drug or activity to produce the same effect that it initially produced. Tolerance is often thought to be a purely physical phenomenon, and tolerance to a particular drug certainly does represent a physical habituation to a substance; however, some studies have shown that there are psychological effects to a person's tolerance for a drug. For example, some studies have administered liquids to participants who were told that the liquid contained alcohol, when in fact it did not, and in many of these studies, the participants began acting as if they were inebriated.

Withdrawal refers to physical and psychological symptoms, usually negative, that occur when the body eliminates the drug from its system after a person stops using it. In the case of some drugs, such as alcohol, the withdrawal process can be life threatening. Withdrawal is often thought to be purely physical, but there have been studies that indicate that drug-dependent subjects experienced different withdrawal severities in different contexts.

Exhibiting tolerance and withdrawal indicates a physical dependence on a substance. This physical addiction is mediated by changes in the body and the brain that make one "dependent" on using the substance in order to feel functional; however, neither are required for a formal diagnosis of addiction. Tolerance and withdrawal do not occur in all addictions.

The Reward System

The study and treatment of addiction goes back hundreds of years; however, the current conceptualizations concerning addictive behavior got their start in the 1950s. A seminal study by James Olds and Peter Milner in 1954 determined that rodents would learn to continuously press a lever in response to having a specific area of the brain electrically stimulated via electrodes implanted in the brain. However, the researchers found that the rats would only learn to press a lever in response to stimulation in an area known as the *septal area* of the brain; stimulation to other areas of the brain did not result in lever pressing.

The septal area is basically the gray matter structures of the thalamic area and the limbic system. These studies, and many others like them, have come to be known as intracranial self-stimulation studies, and the general areas involved in these types of studies are often referred to as *the pleasure centers of the brain*.

The early studies of intracranial self-stimulation were performed with the assumption that the learned lever-pressing behaviors in rats for self-stimulation were different than the lever-pressing behaviors for food rewards and other natural reinforcers. However, subsequent research has indicated that the two brain mechanisms are essentially the same for both. Current thought in addiction physiology has taken the stance that the reward system in the brain functions in addictive behavior in a very similar manner as in normal reinforcement behavior and that the brain mechanisms involved in these behaviors are essentially the same.

Brain Pathways Involved in Addiction

The major brain pathways of the reward system are the mesolimbic and mesocortical dopamine pathways. The mesolimbic pathway begins in the midbrain (in an area called the ventral tegmentum, abbreviated VTA), connects to the limbic system by way of the nucleus accumbens, the amygdala, and the hippocampus, and then projects to the prefrontal cortex. The mesocortical pathway connects the VTA of the midbrain to the cerebral cortex and has especially strong connections to the frontal lobes. These two pathways are two of the four brain pathways that use dopamine as their primary neurotransmitter.

The mesolimbic pathway is commonly believed to be the "reward" pathway (although this notion is certainly not unanimously accepted). The mesocortical pathway is believed to be involved in emotional responses and motivation. The VTA is the primary release site for dopamine in the brain.

> The neurotransmitter dopamine was the first neurotransmitter implicated in addictive behaviors and most likely plays a role in all types of addictions due to its role in reinforcement. However, it is not the only neurotransmitter involved in addiction.

The two brain pathways implicated in the physiological mechanisms of addiction, the mesolimbic and mesocortical pathways, have extensive connections with areas of the brain that are important in memory (e.g., the hippocampus), emotion and motivation (e.g., the amygdala and the frontal cortex), and goal-directed behavior (e.g., the frontal cortex). Therefore, one could surmise that addiction recruits all of these brain pathways and results in compulsive behaviors that are goal-directed toward self-stimulation.

Neurotransmitters and Addiction

Several neurotransmitters have been identified as contributing to addiction. Some may play roles only in specific addictions, whereas others may have a general role. This section begins with dopamine and then looks at several different drugs and their associated neurotransmitters.

Dopamine

There are several lines of evidence to support the notion that dopamine is a primary neurotransmitter that is involved in nearly all forms of addictive behaviors. For instance, in studies of rodents given drugs that are dopamine antagonists (an antagonist blocks the action of a specific neurotransmitter), the animals stop the self-administration of different addictive drugs. The administration of these drugs also diminishes the reinforcing effects of the food pellets.

Several lines of evidence for the role of dopamine in addiction come from the *nucleus accumbens*, an important brain structure that is part of the dopamine pathway. The nucleus accumbens is a collection of neurons in the ventral striatum area that is part of the basal ganglia. Many neuroimaging studies have found increases in extracellular dopamine levels in the nucleus accumbens following either a natural reinforcer, such as food, electrical brain stimulation, or addictive drugs, such as cocaine. Other studies have also indicated that the increased dopamine levels in the nucleus accumbens are related to the experience of getting a reward as well as the expectation of getting a reward. More recent research has indicated that the dopaminergic neurons in the VTA fire in response to the perceived value of a reward, such that a greater than expected reward was associated with increased neural firing and a less than expected reward was associated with decreased neural firing. When the expected reward was received, there was no change in the firing rate of the dopaminergic neurons in the nucleus accumbens.

Studies indicate that normal dopamine functioning is diminished in chronic addicts; however, when the person uses drugs, these pathways become hyperactive. In animal studies, the animals will self-administer injections of addictive drugs directly into the nucleus accumbens, even though damage to the nucleus accumbens or the VTA blocks this behavior. Thus, dopamine appears to play an important function in both the receiving and the expectation of rewards and in mediating the effects of reinforcement.

Stimulant Abuse

Cocaine and other stimulants, such as caffeine and amphetamines, are drugs that directly affect the dopaminergic brain pathways and in some cases alter them. Other excitatory neurotransmitters, such as glutamate, are affected by stimulant abuse. The overuse of stimulants can lead to a type of drug-induced psychosis that resembles the psychiatric disorder *paranoid schizophrenia*. Early theories of schizophrenia suggested that this disorder was related to an overabundance of dopamine in the brain. The primary mechanisms by which cocaine and related drugs work are through their blocking of dopamine transporter molecules in the presynaptic membrane that function to remove dopamine from the synapse and transfer it back into the neurons after it has performed its functions. Other stimulants increase the release of other neurotransmitters in the synapses.

Alcohol

Alcohol is classified as a depressant drug not because it causes depressed mood, although it can do that in large doses, but because at moderate or high levels, alcohol results in inhibition of neural firing, whereas at low doses it may increase neural firing. This is why when one initially drinks alcohol, one may feel excited and uninhibited, but as one continues to drink alcohol, one experiences slowed reflexes, slurring of words, and other similar effects. Alcohol addiction or alcoholism often produces physical dependence such that the person will experience withdrawal symptoms. Like with any other abused drug, the neurotransmitter dopamine is involved in the reinforcing effects of drinking alcohol; however, a number of other neurotransmitters are also implicated in alcohol abuse. Alcohol appears to be an agonist (facilitates the action) for inhibitory neurotransmitters such as GABA and an antagonist (blocks the action) for excitatory neurotransmitters such as glutamate and serotonin. At moderate or high levels then, alcohol use results in decreased firing of excitatory neurotransmitters and increased firing of inhibitory neurotransmitters.

Heroin, Morphine, and Other Analgesic Drugs

Opiates are drugs that are derived from opium, the dried sap taken from the seeds of the poppy plant. Opiates, like heroin and morphine, are associated with high physical dependence because the brain is already wired for them. Endorphins and enkephalins are neurotransmitters that have their own receptors in the brain, and these receptors also have a high affinity to opiate drugs because the structure of these substances is basically the same. Analgesic drugs are extremely effective in the treatment of pain and other symptoms, like diarrhea and persistent cough, but are highly physically addicting because tolerance to these drugs develops quickly and periods of non-use are likely to result in withdrawal symptoms. Moreover, because they inhibit physical functions, such as breathing and heart rate, they are easily overdosed.

Synthetic opiate analgesics, such as OxyContin, produce the same effect as heroin or morphine. Many long-term detrimental physical effects associated with the use of these drugs appear to be due to issues such as sharing needles or poor hygiene, and not directly to the actions of the drugs themselves.

Tobacco, while actually legal for adults, may be more addictive than heroin. Some estimates suggest that about 70 percent of all people who experiment with smoking cigarettes become addicted, whereas 10 to 30 percent addiction rates are reported for alcohol and heroin, respectively. Tobacco is also responsible for far more deaths than any other drug, although nicotine overdoses are rare.

Tobacco

The major psychoactive ingredient in tobacco is nicotine, although there are some 4,000 other chemicals in cigarettes. Nicotine, like heroin, has a ready-made neural receptor in the brain called the *nicotinic cholinergic receptor*. Tolerance develops quickly to nicotine, and heavy smokers are drug addicts in every sense of the word.

Marijuana

The psychoactive effects of marijuana are largely due to a substance called THC (delta 9-tetrahydrocannabinol). However, marijuana contains over eighty other chemicals called *cannabinoids* similar to THC that are probably also psychoactive. There has been quite some controversy lately as to whether or not marijuana is an addictive drug, and while there are a small group of people that do become addicted to marijuana, marijuana's addiction potential is low compared to other drugs discussed in this section. Marijuana also has some medicinal uses.

> There are two confirmed health hazards of marijuana use: Chronic users can develop respiratory problems, and high doses of marijuana can trigger heart attacks, especially in people who previously suffered a heart attack. Current evidence does not suggest that marijuana causes significant brain damage in adults, although this research is ongoing.

Why Can't They Stop?

There are many biologically based theories and biological approaches to addiction. The early attempts to explain addiction related it to the physiological dependence on drugs such that the physical dependence traps the addicted person in a downward spiral of taking drugs to avoid withdrawal symptoms. This is the classic depiction often seen in the movies or media, where the addict is driven to take drugs to avoid withdrawal. The problem with this notion is that when addicts are detoxified and are no longer experiencing withdrawal effects, there is almost a 100 percent relapse rate if no other treatment is undertaken. A second problem with this notion is that many highly addictive drugs, such as cocaine, are not associated with severe withdrawal symptoms. Finally, the pattern of addiction displayed by many addicts involves alternating periods of drug binging followed by relatively lengthy periods of abstinence. The avoidance of withdrawal symptoms can be a motivator to take a drug, but it appears that this is not the primary motivator for drug abuse and addiction. As a result, other theories have been proposed that do not assume that physical addiction is the driving force in addiction.

One group of theories that has received some prominence focuses on the rewarding effects of the substance. These *positive-incentive* theories assume that craving for a substance is targeted for the expected pleasure-producing effects of the drug. The positive-incentive value refers to the anticipated pleasure associated with a particular action, such as drug taking. This expected value is different than the actual amount of pleasure that is experienced from an action (often referred to as the hedonic value of an action). However, an observation that addicts almost universally agree with is that, with time, the effects of the drug are not as pleasurable as they once were, and these positive-incentive theories cannot explain what causes a person to become an addict as opposed to just a user. The compulsive use exhibited by addicts cannot be explained by these theories. Two major theories have addressed these concerns.

One such theory is the *incentive-sensitization* theory that hypothesizes that it is not the pleasure of taking the drug that is the basis of the addiction; it is the *anticipated pleasure* of taking the drug that drives addiction. At first the drug's positive-incentive value is associated with its pleasurable effects; however, as time goes on, tolerance develops, reducing the actual pleasure received from the drug, and the addict's desire for the drug becomes sensitized. Addicts continue to seek the drug for the anticipated pleasure it may bring and not the actual pleasure itself. It is the *wanting* of the drug that drives addiction in this model, not the liking of the effects of the drug.

The Medical Model

The other theory is based on a medical model of behavior and is often termed the *disease model* of addiction. In this model drugs, such as cocaine, when taken chronically, alter the structure of the brain and interfere with rational decision making in the addict. The alteration of the brain occurs in the reward systems of the brain discussed earlier and associated areas in the frontal cortex that control the ability to terminate a repetitive behavior. Addiction is therefore a physical disease, like cancer.

This particular model and its variations are quite popular in medical circles with psychiatrists, physicians, and many addiction treatment professionals. This model forwards the notion that addicts cannot stop using their own willpower, that they need assistance, and that drugs of addiction rob the addict of the power to choose to quit. This model relies on the high rate of relapse in nearly all addictions as evidence that it is a physical process, and not a cognitive one, that drives addiction. The notion of addiction as a disease will be addressed later in this chapter.

Relapse

Relapse occurs when someone who has been abusing a substance or is addicted to a substance quits taking that substance for a period of time and then begins to take the drug again. One cannot relapse if one is not addicted to or abusing a substance. One of the puzzles in addiction treatment is the repeated observation that getting someone to stop using a drug or substance is not typically the problem; the main problem is preventing him from relapsing. Research has indicated that there are several conditions that will increase the probability that a person will relapse:

1. *Priming* occurs when an addict who has been abstinent for some time starts using the drug again in a smaller dose or with intentions to only "have a little." Typically, the addict believes that he has control over his use and will only use again one time; however, often once he starts using again, even with the intent to use significantly less, he reverts back to his old ways. Many addicts will attest to the fact that the relapse behavior is often worse than the original addictive behavior. Drug priming is so prevalent in addiction that many treatment programs believe that only total abstinence from the drug can control relapse.

2. Related to drug priming is the notion that *environmental cues* contribute to relapse. Environmental cues are such things as places, people, or objects that were associated with drug use in the past. These cues are so strong that they often are themselves priming mechanisms that lead to relapse.

3. *Unclear goals*, or not having a clear understanding of the relapse process associated with priming and cues, can lead a person to think that all he needs to do is stop taking the drug and everything will be fine. Addicts can relapse by trying to use only once or by returning to environments associated with using.

4. *Stress* is another factor associated with relapse. This reinforces the view that substance usage and the development of addiction is an attempt to cope with stressful situations in one's life. Stress often results in an addict returning to familiar escape patterns.

One thing that should be clear from the discussion of factors leading to relapse is that there is a strong association between environment and substance dependence, such that addiction may represent a form of classical conditioning. This suggests that cues in the environment strongly pull for an addict to engage in his addiction, and that the brain pathways developed in addiction are powerful.

What Treatments Really Work?

There are a large number of different treatment programs, self-help books, and other potential solutions for treating addictions. Interestingly, the most studied forms of treatment have just about the same relapse rates over the long-term. Many programs, such as the Alcoholics Anonymous twelve-step program, purport a total lifestyle change for the addict, whereas other programs concentrate on just the addictive behavior itself. One of the observations regarding successful outcomes in addiction treatment is the motivation of the addict. A number of studies have suggested that most addicts, such as cigarette smokers, quit on their own. And while many have occasional relapses, over the long run, they are successful. Any treatment program or any behavioral modification program will not work if the person is not committed to the particular goal in mind.

Secondly, support from friends, family members, and even professional support, such as therapy or groups, can be successful in increasing an addict's motivation to stop. Even twelve-step programs that support the medical model of addiction, but that offer no medical treatment, are successful in cases when the addict is motivated and is provided with education and support. One of the advantages to twelve-step programs is that they do not cost anything (they are funded by voluntary donations), they develop a sense of group unity, and one can attend any day of the week. The members in these programs will often go out of their way to help a suffering individual. Other intervention programs, such as residential programs, offer similar benefits, but are time-limited. Many ex-addicts believe that addiction recovery is a lifetime endeavor and requires some form of ongoing support for a longer time period.

Taking Drugs to Stop Drugs

There are some medications that have demonstrated some success in treating people with certain types of addictions. By far the most commonly used of these is a medication called *Antabuse*. Antabuse interferes with the breakdown of alcohol in the person's system and results in the person becoming violently ill if she drinks alcohol while taking the drug. Other drugs, such as Revia or Suboxone, were designed to treat narcotic addictions, but have also been used to treat alcoholism and addictions to other drugs and to help eliminate cravings. Most studies have found that these drugs are only moderately effective because they can only work if the person takes them. Thus, any medications prescribed to halt addictive behaviors are only as good as the addict's own commitment to quit.

Is Addiction a Disease?

There are several things that are clear regarding addictions in humans. First, addiction is a psychologically complex phenomenon. Second, there has been much evidence to support the notion that addiction is a disturbance of decision making and that addicts have poorer decision-making skills than do normal controls. Third, addiction is not limited to just drugs but includes many other behaviors, including gambling, shopping, eating, sexual activity, and many others. Fourth, addiction, like many behaviors, has a strong genetic component in that a person's potential for becoming an addict is increased if first-degree relatives have had an addiction. Finally, the change from the initial use to compulsive addiction is associated with structural changes in the limbic system and frontal cortex and involves many neurotransmitters.

Those who support models of addiction as a disease point to the fact that these changes in brain structure of addicts are responsible for an individual's inability to stop her addiction. The evidence that they bring to the table is quite impressive and convincing.

However, those who refute the notion of addiction as a disease point out successes of treatment programs, such as twelve-step programs, as evidence that addiction is

> People become committed to quitting substance abuse and addictive behaviors when these behaviors conflict with important goals in their lives or when certain life-changing events occur, such as having a child. Not everyone follows this pattern, but when an addict views his addiction as harmful or limiting his choices, he often takes positive steps. The problem is that most addicts consider their addictive behavior a positive choice.

not a disease. They ask, "If addiction is a physical disease like cancer or heart disease, then how can attending a group, following a set rules, and interacting with others who have the same disease cure the disease?" There are no twelve-step programs that cure cancer, diabetes, or other physical diseases. While addiction certainly has a physical component to it (as do all behaviors), this viewpoint supports the notion of addiction as a volitional behavior.

The answer to this question is not a simple one. If addiction is a brain disease, it certainly is not like other brain diseases, such as Alzheimer's disease, Parkinson's disease, and others, where behavior is not goal-directed but instead becomes erratic and disorganized. Moreover, the proposal that drug use changes the brain should not be all that shocking. The brain changes in response to its environment, whether it occurs due to drug use, learning to play the piano, reading, or any number of activities. The final answer depends on what one accepts as a definition of what constitutes a *disease*. If one accepts the notion of a disease as a social metaphor for dysfunctional behavior, then perhaps addiction is a disease; however, if one accepts the traditional notion of a disease, perhaps the criteria for a true disease are not fully met by addiction. Moreover, there is no evidence that labeling addiction as a disease or believing that it is a disease offers any advantages to recovering addicts. Addicts who believe that addiction is a disease relapse as often as those who do not.

BRAIN-RELATED DISORDERS

MUCH OF THE UNDERSTANDING of how the brain works was learned via those with brain damage or a brain disorder. The brain is more plastic than originally believed, but there is also a level of brain damage from which the brain cannot recover, as seen in certain brain diseases like Alzheimer's disease. A full discussion of all the known brain disorders and psychiatric conditions would cover volumes of books. The goal of this section is to discuss some specific disorders that are of general interest and demonstrate how they relate to impairments in brain functioning and cognition.

Depression and Mood Disorders

Everyone has experienced a depressed mood at one time or another. Feeling down or depressed is not the same thing as having *clinical depression*, which is a psychiatric disorder that is associated with certain recurrent symptoms, just one of which is depressed mood. On the other hand, most people have also had periods where they felt excessively happy or "up." Feeling happy or elated is not the same thing as *mania,* which represents a severe clinical symptom associated with bipolar disorder. The difference between the normal emotional states of being either "up" or "down" and clinical disorders is a matter of both degree and duration. Clinical mood disorders, such as clinical depression and bipolar disorder, represent rather severe, extended, and extreme degrees of normal emotional experiences. Often it takes a seasoned professional to be able to tell the difference between clinical depression and normal depressed mood.

Major Depressive Disorder

Clinical depression, or major depressive disorder, is a mood or *affective disorder* that is one of the most commonly occurring psychiatric disorders in the United States. An estimated 15 to 20 percent of all Americans experience clinically significant depression at one point in their lives. Depression can occur across any demographic group or at any stage of life, but it is diagnosed about twice as often in women, and the peak frequency years for depression are in the twenty-five to forty-four-year-old age range. Depression also tends to run in families.

The common symptom in major depressive disorder is, of course, depressed mood. People with depression also suffer a number of other cognitive and physical symptoms, such as anhedonia (the inability to experience pleasure), pessimism concerning the future, disrupted patterns of eating and sleeping, hopelessness, cognitive and motor slowing, anxiety, and even suicidal tendencies. A specific number of symptoms must occur together for a period of at least two weeks before a person can be diagnosed with clinical depression. In addition, clinical depression can be mild, moderate, or severe in its expression. Some earlier viewpoints of depression considered it to be either *reactive* or *endogenous* in nature. *Reactive depression* develops in response to negative life experiences, whereas *endogenous depression* has no such apparent cause. This designation is not used often today in clinical usage but is still seen in textbooks.

There have been many neuroimaging studies of patients with depression and with bipolar disorder. In many of these studies, differences in several brain structures have been reported, but the pattern of results has been inconsistent. In some individuals who have chronic, severe depression, smaller cortical volumes were observed. Additionally, two structures have been repeatedly found to be abnormal in many structural and functional neuroimaging studies of people with mood disorders: the amygdala and the anterior cingulate cortex, both of which are involved in memory and emotional regulation. With respect to what causes clinical depression, there are two major theories that remain popular in the clinical literature.

Billions of dollars are spent each year on antidepressants. Numerous studies evaluating the effectiveness of these drugs have found that the three classes of antidepressants typically produce equivalent results. One large study found that 50 percent of clinically depressed patients improved using these medications; however, the no-treatment control group also showed a 25 percent rate of improvement. Does this suggest that antidepressant medications are effective?

The first theory applied to explain the cause of many psychiatric disorders is the *diathesis-stress model*. This model purports that certain people are born with an inherent or genetic susceptibility (a diathesis), which by itself is incapable of causing a psychiatric disorder, but interacts with stress to produce a specific psychiatric disorder.

A second theory to explain the causes of depression is the *monoamine theory of depression*. This theory is largely based on the observation that antidepressant medications appear to be effective in treating depression. Monoamines are a group of neurotransmitters that are related based on their chemical structure, and include serotonin, norepinephrine, and dopamine. The evidence for the validity of this theory is sparse, but there have been studies examining the brain tissue of deceased clinically depressed individuals who were not treated with antidepressant medications, and deceased normal controls. There has been some evidence from these studies, again not consistent, that the depressed patients had more serotonin and norepinephrine receptors in their brains than did the controls. This observation is believed to be due to a biological process called *up-regulation*. Up-regulation occurs when an insufficient amount of the neurotransmitter is released into the synapse, leading to a compensatory increase in the number of receptors for that specific neurotransmitter. However, the studies are not consistent in their findings, and the treatment rate for antidepressant drugs is not any more effective than treatment with psychotherapy, exercise, or other forms of treatment.

Bipolar Disorder

Bipolar disorder represents an alteration between periods of severe clinical depression and shorter periods of mania. Mania is characterized by extremely increased activity, high levels of energy, restless activity, rambling speech, and the loss of inhibitions. Manic people can often pose a danger to themselves and others. Unlike major depressive disorder, bipolar disorder is not common and occurs in 1 to 4 percent of the population.

> *Unipolar depression* is another term for major depression and is aptly named because the depressive symptoms typically occur in only one direction, down or sad. Bipolar disorder was once known as manic-depressive disorder, but received the new label as the mood in this disorder is bidirectional, alternating from mania to depression.

The results of neuroimaging studies indicate that people with bipolar disorder tend to have a larger amygdala. There also appears to be a genetic association for bipolar disorder in that an identical twin or first-degree relatives of a person with this disorder are at increased risk to develop it. However, the exact genes responsible for this disorder have not been found.

The treatment for bipolar disorder includes lithium and mood stabilizer drugs that are often used in the treatment of epilepsy. How these drugs work is not yet known, but it is thought that they affect certain excitatory messenger systems in the brain.

Seasonal Affective Disorder

Seasonal affective disorder, appropriately labeled SAD, is comprised of regular bouts of depression that reoccur during a particular season, such as winter (winter is the most common season for SAD to occur, but it does occur in other seasons). SAD is more prevalent in regions where winter nights are long but also occurs in moderate climates. There are several treatments for SAD, which include exposure to ultraviolet light for an hour a day, antidepressant medications, and psychotherapy.

Schizophrenia

Schizophrenia is a heterogeneous disorder characterized by any of the following symptoms: intellectual deterioration, emotional blunting, disorganized speech, disorganized behavior, social isolation, delusions, and/or hallucinations. Schizophrenia has a prevalence rate of approximately 1 in 100 persons. It is one of the most crippling mental disorders and at the same time one of the least understood. Schizophrenia is not a gender-specific or culture-specific disorder; it exists in all cultures, and the ratio of overall cases favors neither men nor women. The disorder most often appears before the age of twenty in men, whereas women often have a later onset with more mood disturbance and a better prognosis. Schizophrenia accounts for nearly one-third of the homeless population in the United States and costs the United States alone over $50 billion a year.

Current Diagnostic Issues with Schizophrenia

Schizophrenia consists of five subtypes. These subtypes are defined based on the presence of positive symptoms (excesses) or negative symptoms (deficits) of behavior in the presentation of the disorder. Positive symptoms are psychotic behaviors (a loss of touch with what is real). The positive symptoms in schizophrenia include:

1. *Hallucinations* are sensory distortions. Often the person will hear, see, smell, or feel things that others do not. Auditory hallucinations are the most common type of hallucination in schizophrenia, and these most often consist of hearing voices that no one else can hear. These voices might talk to the person, tell him things, order the person, talk to each other, etc. Other common types of schizophrenic hallucinations include seeing people or objects, smelling odors, or feeling things, such as bugs crawling on the skin or being touched by someone or something that is not there.

2. *Delusions* are very strong "false" beliefs. Often the person will still adhere to the belief even after being shown that it is not true or illogical. People with schizophrenia will often express seemingly bizarre beliefs, such as believing that aliens or the government can control their behavior through a television or other means. Other common delusions include the belief that people on television or the radio are directing messages to them or are broadcasting their thoughts aloud to others. Paranoid delusions or delusions of persecution include the belief that others are trying to harm them, cheat them, spy on them, harass them, or plot against them. Delusions of grandeur include the belief that one is an extremely important person like a messiah or that one is a famous historical figure.

3. *Thought disorders* are dysfunctional ways of thinking. Several common forms of thought disorders are often seen in people with schizophrenia. Disorganized thinking occurs when the person cannot organize his thoughts or connect them in a logical and meaningful manner (sometimes called a flight of ideas). Thought disordered individuals often make up meaningless words (called neologisms).

Thought blocking occurs when a person stops speaking abruptly in the middle of a sentence or in a specific train of thought. When asked why he stopped, the person will often report that the thought had been removed from his head.

4. *Movement disorders* can present as disorganized or repetitive agitated body movements. At the other end of the spectrum, catatonia occurs when the person does not move and does not respond. People may spend prolonged periods of time in odd stances.

Negative symptoms consist of disruptions to one's emotions and motivations. Often these types of symptoms are mistaken for depression or other conditions. Negative symptoms include:

1. *Flat affect*: a person does not express emotion, or she talks in a dull or monotone voice.

2. *Anhedonia*: an inability to experience pleasure in one's activities.

3. *Amotivation*: an inability to begin, plan, and sustain activities.

4. *Poverty of thought*: observed as a poverty of speech, which is speaking very little or being mute even when the person is forced to interact with others.

5. People with negative symptoms need help to complete everyday tasks and will often neglect basic personal hygiene. This may make them appear lazy or unwilling to help themselves, but these problems are due to the symptoms of schizophrenia. Cognitive issues in schizophrenia have to do with attention, executive functions, working memory, and other problems attributed to the frontal lobes of the brain. There are five subtypes of schizophrenia: paranoid, disorganized, catatonic, undifferentiated, and residual.

Schizophrenia is not a disorder of multiple personalities, as it is often depicted in popular media (that disorder is called **dissociative identity disorder**). The term was coined by the psychiatrist Eugene Bleuler in the early 1900s and means "split mind" (schizo = **split**; phrenia = **minds** in Greek). This was depicted by Bleuler as a splitting of cognition from personality.

The Cause of Schizophrenia

There is no defined cause for any form of schizophrenia, although many have been proposed. Early in the conceptualization of the disorder, bad or inconsistent parenting and influenza in the mother during pregnancy were offered as possible causal factors. While today the field of psychiatry has denounced causal explanations of schizophrenia as being related to child-rearing practices or early experiences of the child, the influenza explanation still receives some support. However, today modern psychiatry has pronounced schizophrenia to be a "disease" that is biologically based (the typical psychiatric explanation for many mental states).

Two major factors are commonly believed to contribute to the cause of schizophrenia: First, it is generally acknowledged that schizophrenia is at least in part caused by an imbalance of neurotransmitters. The classical "dopamine hypothesis" of schizophrenia has asserted that there is a hyperactivity in dopaminergic transmission at the dopamine D2 receptor (a specific type of dopamine receptor in the brain) in the connections to the limbic system in the brain. Despite several limitations, this hypothesis still remains the most popular of the neurochemical theories. The idea that dopamine is central in schizophrenia is a notion that was initially supported by the observation that dopamine agonists (substances that increase dopamine in the brain), such as cocaine and amphetamines, can induce psychosis in healthy subjects and aggravate psychotic symptoms in schizophrenics. Several receptors have been implicated in schizophrenia; however, it has been suggested that up-regulation of D2 receptors could be the result of adaptation to antipsychotic medications. Some PET studies have demonstrated no significant difference in D2 receptors between never treated schizophrenics and a control group. More recent evidence has indicated that other neurotransmitters are also affected in schizophrenics, such as glutamate and serotonin. Of course, much of this evidence has come from the recent development of the atypical neuroleptics (a type of antipsychotic drug), whose main mechanism of action is not just the blocking of dopamine.

There is six times the prevalence of schizophrenia in the homeless population compared to the general population. Contrary to the depiction in the media, paranoid schizophrenics or people with other forms of schizophrenia are not typically serial killers or violent, dangerous criminals. In fact, the crimes they most often commit are vagrancy and trespassing.

The other line of evidence is heredity, suggesting genes play a role in the cause of schizophrenia. However, it is still unclear if schizophrenia is the result of a single mutated gene, a series of mutated genes, or a mutated gene passed from parents. The concordance rate between identical twins (the probability that one twin will have the disorder if the other has it) for schizophrenia is around 0.5 in most studies. Several groups of researchers have hypothesized that different genotypes are responsible for different manifestations of schizophrenia.

Schizophrenia has been shown to affect both cortical and subcortical brain tracts, especially in the anterior region of the brain. Gene mutations may explain why the brains of schizophrenics have larger ventricles and less gray matter; however, some researchers note that this finding is only consistent in schizophrenics treated for long periods with medications. Still, concordance rates are not 100 percent, so there appears to be more than just a genetic basis for it.

Autism

Autism is a complex developmental disorder that is classified as a *pervasive developmental disorder*. A pervasive developmental disorder consists of a spectrum of disorders that typically emerge in early childhood and affect a large number of cognitive and physical domains. Autism is itself a heterogeneous disorder because it consists of a spectrum of presentations, ranging from severely cognitively impaired to functional. In general, most children diagnosed with autism tend to have severe deficits in social functioning and with their verbal abilities.

In almost all cases of autism, the disorder manifests itself at an early age, typically before the age of three years old. There are core symptoms of autism that are displayed in some form, but not all need to be present in the same person. These core symptoms include:

1. Deficits in the ability to understand the intentions of others and to interpret emotions that other people display. For example, an autistic child will not smile back when you smile at him and will typically not be able to discern nonverbal communication, such as voice inflection.

2. An extremely reduced capacity for communication and social interaction. Typically, the first thing parents notice in these children is their lack of a response to the parents' attempts to reciprocate emotions in them, such as happiness, smiling, cooing, etc. They are very distant and detached. Moreover, it is common for autistic children to display delayed development in verbal abilities, such as talking.

3. There is a preoccupation with a single subject or activity that is highly irregular from the typical preoccupations that children display.

Four-fifths of children diagnosed with autism are male; one-third have seizure disorders; and nearly one-half suffer from some level of mental retardation, ranging from mild to profound. The probability of having an autistic child increases if either parent is over forty years of age. Another interesting observation about autism is that in the early 1900s the incidence of autistic births was estimated to be less than 1 in 1,000. Currently, estimates of the incidence of autism are between 1 in 160 to 1 in 330 births, a dramatic rise. This has led to speculation that certain aspects of modern technologies, such as food preservatives, vaccinations, and other variables, are responsible for the increase in autism; however, there also has been a drastic shift in the diagnostic criteria such that the disorder has much more liberal diagnostic criteria now than it did in the early 1900s. This alone could account for an increase in diagnoses of autism, along with better public awareness and improved communications about childhood disorders. However, there is reason to believe that the increases in diagnoses of autism have risen despite these issues. There certainly is a cause or multiple causes to autism, and none of these can be fully ruled out.

> One myth about people with autism is that they all have "special" abilities. Less than 10 percent of children with autism display savant behavior. There are cases where autistic savants have amazing abilities, like being able to report dates of any historical event, performing amazing feats of arithmetic and estimation, and others. However, these savants have pervasive deficits elsewhere and cannot learn to use their gifts constructively.

There is a genetic contribution to any neurological disorder, and autism is no exception. The findings suggest that autism may be triggered by the interaction of many genes interacting with the environment as opposed to the action of a single genetic cause. Many autistic children can perform well on tests involving rote memory, music, art, and other activities that are not highly verbal in nature, whereas verbal activities tend to be adversely affected. Numerous structural imaging studies and postmortem studies of autistic patients suggest there are several areas of brain pathology, including the cerebellum, the frontal cortex, and the limbic cortex. One line of research attempting to determine the cause of autism focuses on mirror neurons, neurons that fire when someone observes or imitates another performing an act. It has been hypothesized that there are deficits in this system and that autism is a deficit in the *Theory of Mind*, or in the ability to develop personal theories of how other people think and feel. However, there is no overwhelming evidence that any particular deficit results in autism.

Dementia

Dementia literally means "without mind." Dementia is a general term for a number of brain disorders and other conditions that result in significant intellectual impairment. There are over fifty different conditions that are associated with dementia; therefore dementia is not a specific disease or specific condition, but instead, a term that describes this decline in cognitive functions. Many conditions that can cause dementia are treatable, such as severe depression. Other conditions, such as Alzheimer's disease or Huntington's disease, have no known cure. It would be impossible to list *all* of the known causes of dementia in this chapter; however, several causes of dementia are discussed in this section.

> There are several different classifications of dementia. Dementia can be classified as to whether it primarily affects the brain's cortex or subcortex (*cortical* versus *subcortical* dementia); as a *progressive dementia* that steadily worsens over time; or, dementia can be classified as either *primary* (not caused by any other disease or condition) or *secondary* (results from another condition).

Alzheimer's Disease

Alzheimer's disease (AD) is the most common form of dementia. AD typically begins with short-term memory loss followed by more serious memory loss. Then the individual will experience deficits in other cognitive domains, such as language or executive functioning. The disease can progress slowly or rather rapidly, with a range of progression from one year to as many as twenty years. The average length of the disease is about eight years. AD is always fatal, and despite several medications on the market, there is no cure. AD is a primary, progressive, cortical dementia.

While there is no way to predict who will get AD, several risk factors have been identified. As one ages, the risk doubles for every five-year increase after the age of sixty-five. High blood pressure and diabetes are significant risk factors. Genetic risk factors, such as being positive for the APOE gene, have also been identified, but these only offer mild predictive value.

AD is characterized by specific brain pathologies: There is massive neuronal death and the development of amyloid plaques and neurofibrillary tangles. Amyloid plaques, found in the tissues between the neurons in the brain, are made of a protein called beta amyloid as well as degenerating neurons and glial cells. Neurofibrillary tangles are bundles of twisted filaments found within neurons. Neurofilaments are the support structures of the neuron, the cytoskeleton. The neurofibrillary tangles are largely composed of *tau proteins*. In a healthy neuron, tau protein aids in the functioning of microtubules, which transport nutrients and other substances throughout the neuron. In AD, tau is altered so it twists into pairs of helical filaments that become tangled, thus inhibiting normal cellular functioning. As neurons die, the brain connections also die. This process initially occurs in the medial temporal lobes where the hippocampus is located, hence the initial symptom of an inability to form new memories, but as the disease progresses, this process occurs throughout the brain.

Medications for AD were developed based on observations that the areas of the brain that use the neurotransmitter acetylcholine were affected early in the course of the disease. So medications, such as Aricept, were developed to stop the breakdown of acetylcholine in the brain. These medications do not stop the progression, but in some studies have been shown to slow the progression of AD. Other newer medications are glutamate antagonists that slow down neuronal death due to excitotoxicity, which occurs when neurons fire in prolonged bursts due to being excited by glutamate, resulting in the neurons becoming damaged and eventually dying. This situation occurs with neurochemical and

It is not true that severe memory loss, as occurs in dementia, is associated with the normal aging process. Studies following participants over time have found that mild deficits in memory did not occur until the seventh decade of life. Dementia is **not** a result of normal aging.

neuropathological changes involved in neurodegenerative diseases, such as AD and other forms of brain injury. Medications, such as Namenda, slow this process down. However, AD is always fatal, and these medications can only slow down the progression.

Lewy Body Dementia

Lewy body dementia (LBD) is another common type of progressive dementia. LBD often occurs in people with no known family history of the disease, whereas AD has demonstrated a strong hereditary association. In LBD, cell loss occurs in the cortex and in the substantia nigra (recall this structure from the discussion on Parkinson's disease), and there is also the formation of Lewy bodies, the hallmark of LBD. Lewy bodies appear in the substantia nigra and in the cortex, and contain a protein called alpha-synuclein that has been linked to Parkinson's disease and several other disorders. At this time, it is unclear why this protein accumulates in the brains of these individuals, but some Lewy body formation is also found in the brains of many normal controls.

The symptoms of LBD appear similar to AD; however, those with LBD typically will also experience visual hallucinations and Parkinsonian symptoms, such as a shuffling gait, rigid posture, and fluctuations in the severity of symptoms.

Vascular Dementia

Like any organ, the brain receives nutrients and oxygen from blood delivered by a sophisticated vascular structure. *Vascular dementia* was once known as multi-infarct dementia and occurs when vascular changes in the brain result in a loss of oxygen to the associated brain area, leading to neuronal death. These *vascular infarcts*, which are dead pockets of tissue scattered in various areas of the brain, can lead to a "patchy" presentation on neuroimaging studies, because these infarcts typically occur in scattered areas throughout the brain. Because of this process, people who have significant vascular problems can exhibit a variety of cognitive deficits, such as difficulties with language, forgetfulness, motor problems, and other problems.

There are many potential causes of vascular dementia, such as stroke, vasculitis (an inflammation of blood vessels), and the autoimmune disease lupus erythematosus. Binswanger's disease, a rare form of vascular dementia, is characterized by damage to small blood vessels in the white matter of the brain.

Vascular dementia follows a different course than does AD; where AD is progressive and manifests in a downward decline, vascular dementia progresses in a more stepwise decline. The individual will seem to be functioning at one level for a period of time, and then there will be a sudden drop in her functioning, followed by another plateau (but at this new lower level of functioning), followed by another sudden drop, and so forth. If the cause of the vascular problems can be addressed, it follows that vascular dementia should be able to be somewhat controlled; however, any damage to the brain that had already occurred would not be expected to remit.

Stroke

The term "stroke" was used to describe the sudden onset and devastating effects of what is now technically termed a *cerebrovascular accident* (CVA). It appeared that the individual was struck down by God when he experienced a stroke (hence the name). CVAs occur as a result of some type of disruption or blockage in the vascular system of the brain. CVAs can be classified in several different manners:

A *hemorrhagic CVA* occurs when the vessels rupture and blood leaks into the brain, whereas an *ischemic CVA* is one where the vessels are blocked and blood flow is restricted or reduced in a specific brain area, causing that area of the brain to deteriorate due to lack of oxygen and nutrients. The outcome is typically better for hemorrhagic CVAs, as ischemic strokes often represent chronic damage.

An *embolic CVA* occurs when a clot travels through the bloodstream and lodges itself in the vasculature of the brain, whereas a *thrombotic CVA* occurs when the blockage develops at a particular site in the vasculature of the brain.

CVAs can also be classified by their location, such as the particular artery, the particular area of the effects, and others. An *aneurysm* refers to a "bubble" that develops in the vessel or artery, and blood flow becomes restricted or sluggish due to this occurrence. The development of an aneurysm appears to be related to both genetic and environmental effects, and aneurysms can burst and result in hemorrhagic CVAs.

A *lacunar infarction* typically occurs when small blood vessels are blocked and unable to feed the tissue, resulting in that area of the brain dying and deteriorating. The term comes from the Latin word *lacuna* meaning "hole," which is what a lacunar infarction resembles, a small hole in the brain.

A *transient ischemic attack* (TIA) is often referred to by laypersons as a "silent" or "mini stroke" and occurs when the blood flow to the brain is interrupted in a particular area. The effects last for twenty-four hours or less. These effects are typically mild but can mimic a full-blown CVA. The occurrence of TIAs should be taken as a warning sign.

The effects of strokes can be variable, and different people with the same type of CVA can have extremely different presentations. In addition, the recovery from stroke is also quite variable and depends on a number of factors, including the person's age, gender, state of health, motivation, how much therapy or treatment he or she gets, and of course, the severity and location of the CVA. There are a number of risk factors that are associated with CVAs, the most important of which is a prior history of a CVA. People who have had strokes in the past are at a much higher risk to have a stroke in the future. This means that additional strokes will continue to damage the already damaged brain and most likely result in further deterioration. Of course there are other risk factors including smoking, alcohol use, heart disease, diabetes, as well as other risk factors for vascular disease.

In general, the effects of a CVA will depend on its location in the brain. A left hemisphere CVA will often result in difficulties with right-sided motor functioning, including paralysis or severe weakness, language problems, problems with recall for verbal material, and mood issues. A right hemisphere CVA will typically result in left-sided body weakness or paralysis, problems with attention, problems with spatial abilities, difficulties with organization, and other effects. Strokes occurring in the occipital area of the brain will often disrupt vision, whereas strokes occurring in the frontal portion of the brain can be quite variable in their presentation and can include any or all of the aforementioned symptoms. Massive CVAs involve more than one area of the brain and can be quite devastating. Strokes can also occur in the midbrain, the subcortical regions, and in the brain stem, and these may have more generalized and severe effects. Recovery is generally better the more quickly the stroke is treated. The occurrence of multiple strokes in the brain, or even a single severe stroke, can lead to vascular dementia.

SOME BRAIN MYTHS
That You May Still Believe

INCREDIBLY, EVEN IN A SOCIETY where information is easily accessible, people still believe many of the most implausible things imaginable, such as the reliability of psychics or that the brain adds wrinkles when you learn something. Other myths about the brain are no less rare. Here are several popular ones.

You Only Use 10 Percent of Your Brain

At one time or another, every reader has undoubtedly heard the assertion that humans only use 10 percent (or some other small percentage) of their brains. No matter how many times this myth is debunked, it still seems to be popular. The origin of this particular myth is not certain, and many have tried to trace its beginnings back to the early American psychologist William James. James, along with a colleague, had postulated that someone might be able to summon up mental energy when needed in the same way an athlete often summons up a "second wind." James was not referring to some hidden brain power of mystical proportions that people do not normally utilize. Nonetheless, what seems to have happened is that bestselling author Lowell Thomas misinterpreted James's notion to mean that people only use 10 percent of their brains and included it in the forward section of Dale Carnegie's bestselling book *How to Win Friends & Influence People*. The notion that people use only 10 percent of their brains has been propagated by this and by a number of other sources. In fact, there has never been any scientific evidence that people use only a small percentage of their brains, and many scientific publications have gone out of their way to renounce this myth. The myth has become so prevalent that it has been attributed to famous scientists, such as Albert Einstein, and propagated by numerous misinterpretations of academic terms. In fact, no one has ever found a reference by Einstein to this idea.

The "10 percent myth" has been used to justify a number of dubious self-help programs, lay theories, and explanations of brain potential. There are a number of self-help programs and psychic phenomena training programs that promise to help the individual "wake up" the dormant part of her brain and realize amazing "hidden powers." Moreover, certain parties have misinterpreted the results of scientific studies of individuals who were born with a significant portion of their brain missing, but who still demonstrated fairly normal development, to support the notion that people only use a small portion of their brains. Recall that the brain is plastic; at an early age the brain is quite malleable and can make adjustments to the environment. Individuals who were born without one cerebral hemisphere, or who incurred significant brain damage at an early age, have learned to use language and other cognitive abilities and on the surface appear fairly normal, but when they undergo extensive testing, significant deficits are discovered. Just because the plastic human brain can correct for a number of significant injuries at a young age does not mean that people with normal brains only use a small proportion of their brains.

From a scientific standpoint, there are several disconfirming points of evidence that are important to consider when discussing the 10 percent myth:

The observation that "idiot savants" exist has been used to bolster the 10 percent myth. Nothing could be further from the truth. Such savants are typically severely dysfunctional in almost all other areas and, most of the time, cannot use their "gift" in any functional manner.

First, the organs in the body are not wasteful. Your heart does not normally work at a 10 percent capacity unless it is somehow injured or damaged. Neither does any other organ in your body. Why would the brain naturally function below its capacity? The brain comprises a very small proportion of body weight (about 2–3 percent), and yet it uses one-fifth of the body's oxygen and gets first exposure to all nutrients (which are converted into glucose in order to transfer across the blood-brain barrier). Think about this: If the organ that uses one-fifth of the body's intake of oxygen is only operating at 10 percent efficiency, then increasing its efficiency to full capacity would use up all the oxygen in the body. All other organs and tissues would die of suffocation.

Second, patients who are in vegetative states typically have much more than 10 percent of their brains intact. If a person used only 10 percent of his brain, then a person could sustain significant brain damage and still be functional; however, this is never the case, and even patients with severe dementia have significantly more than 10 percent of their brain left.

Third, thousands and thousands of neural imaging studies have indicated that a person is using a significant portion of his brain when processing information. In fact, most neuroimaging studies indicate that a significant portion of the brain is activated when you are doing nothing at all. There is never a time, outside of having severe brain damage, when a significant portion of your brain is not activated.

If one considers the 10 percent myth to be a metaphor, it has some plausibility, but only as a metaphor for the amount of effort, planning, and motivation that people direct toward most tasks that they perform. However, saying that people can be more motivated or have better planning skills is quite a different thing than saying that the brain is only operating at 10 percent capacity. Outcomes in life are dependent on a number of variables, including personality variables, intelligence, serendipity, preparation, and plain hard work. Certainly, people can put more effort or motivation into a good deal of their efforts and can always learn new skills. However, the 10 percent myth is just that, a myth.

Repressed Memories

This topic is a bit more controversial than the preceding one. The notion of repressed memories suggests that a very traumatic experience can be repressed or pushed out of the depths of the conscious mind deep into the unconscious mind, where it is essentially forgotten. The repressed memory then surfaces as depression or anxiety without a known cause and can be uncovered by professional help. The notion of repression was popularized by Freud, and in recent times, the issue of repressed memories has been highlighted in criminal investigations, such as child abuse and even some murder cases. Many psychologists and psychiatrists strongly doubt the existence of such memories; however, others are convinced that repressed memories occur.

Part of the issue with repressed memories is the nature of memory itself. Your memory is not like a tape recorder. When you remember something, you do not remember it exactly as it happened. In fact, when you remember events from your past, you actually recreate them in your mind. Research has shown that recall of past events, even events that are not in the distant past, can be affected by a number of things, such as the context in which they are recalled, the type of questions asked, and even trivial factors like a person's mood or physical state. The issue with repressed memories, if they do exist, is whether they give an accurate indication of past events (probably not). There are myriad reasons to believe that repressed memories do not occur.

First, post-traumatic stress disorder (PTSD) occurs when a severely stressful event results in a severe reaction in a person. This physical reaction is never one of repression, but consists of reliving the event through flashbacks and/or dreams and avoidance of stimuli that remind the person of that event. Memory for certain aspects of the event can be clouded due to the trauma, but the memory is never repressed; it is often relived over and over. This disorder alone suggests that the idea that severely traumatic events are repressed into the unconscious mind is not a probable occurrence. If repression occurs, then why don't PTSD patients repress these traumatic experiences?

Secondly, the method by which most of these "repressed memories" are uncovered is questionable at best. The most popular method is known as "repressed memory therapy." This occurs when a therapist uncovers repressed memories through a series of questions. Typically, when these therapy sessions are examined, the therapist asks a lot of leading questions to a very gullible client, and the so-called memories that are extracted from this process are dubious. Other issues regarding the unreliability of memory and the susceptibility of memory distortions to context during this method of questioning add to the notion that repressed memory therapies are not valid. Studies of memory distortions have

Related research by Elizabeth Loftus has indicated that memories for events that never happened can be strategically implanted in children and in adults. Often, once implanted, these "false memories" are believed to be real, and the person will often swear that the event happened, even though the researchers have proof that it did not.

indicated that the "memories" recovered under these conditions are highly unreliable and often not verified.

Third, in many of the well-publicized cases where people have claimed to have recovered repressed memories, the facts have not aligned very well with the so-called memories of the events when subjected to reliable collaboration with third-party sources.

Nonetheless, people do repress things all the time. The issue is whether you can repress a very traumatic event for years and then have it recovered. The bulk of the evidence indicates that in most cases these recovered repressed memories are not recollections of actual events, but are confabulations. However, theoretically it is true that repression as a defense mechanism is used by many people to avoid unpleasant thoughts while concentrating on other life issues. But these people don't forget these events for years and act as if they never happened; eventually they deal with them on their own.

Subliminal Persuasion

There is a large market for subliminal CDs that make a number of claims about their ability to improve everything from memory to grades in school to quitting smoking to forcing people to act in a certain way. There are still claims that the government, rock bands, and special interest groups are influencing people subliminally, and bending them to their wills. Google the term "subliminal CDs" and over a million hits occur.

Subliminal literally means "under the threshold" and is derived from the "limen of consciousness." In psychology, the *limen* refers to the sensory threshold where a stimulus goes from being just detectable to just undetectable. Stimuli far below the detection threshold cannot affect behavior, whereas stimuli between detection and recognition might influence behavior, but never at the extent claimed by subliminal marketers.

For instance, priming is a well-known psychological phenomenon. Primes are stimuli presented at such brief periods that the person cannot consciously recognize them, but they can be processed by the brain. Primes have been demonstrated to affect the speed and accuracy that people can later identify the stimulus or similar stimuli, respond to a task, or affect someone's mood. For instance, a famous study had students generate potential ideas for research projects. Then the students were exposed either to primes of pictures of a smiling friend or a photo of their angry faculty supervisor. Those seeing the friend rated their ideas as favorable, and those seeing their supervisor rated their ideas less favorably. Such effects do not last very long and never move a person's behavior in a direction that would be out of character for her. A "spreading activation" mechanism, where the prime activates related concepts or emotional states in the brain, explains how primes work. Primes are used in studies to describe differences in cognitive processing in brain-damaged people or to understand how certain cognitive processes work.

However, the claims for subliminal messages stating they can make people buy certain products, improve memory, or indoctrinate people into Satanism have been refuted numerous times. All of these claims appear to be influenced by now-discarded Freudian views of the unconscious mind, outside of awareness, directing behavior. People perform activities and mentally process things outside of their immediate awareness, such as driving a car or multitasking, but people are not slaves to subliminal influences.

The Judas Priest trial in the 1990s occurred when the families of two boys who committed suicide claimed that a song by the band had told the boys to commit suicide, a case of subliminal persuasion. However, after all the evidence was presented, the ruling indicated that there was no scientific evidence that this was possible.

The subliminal persuasion market appears to have begun based on a book by Vance Packard titled *The Hidden Persuaders* back in the 1950s. Packard described the claims of a movie theater owner who stated that he was able to increase concession sales after flashing images on a movie screen for 1/3000 of a second (way too fast to be recognized). Even though in 1962 this same person admitted that he made up the whole story, all kinds of subliminal improvement programs followed the release of the book. In addition, this book led to claims that rock bands included hidden satanic messages in their lyrics that were exposed when playing the music backward and that corporations like Disney have included subliminal messages in their films. However, if you cannot understand a backward message in a song because you cannot recognize it when the music is played forward, how can it influence you subliminally? Many research studies have totally refuted the notion that any such messages influence behavior. For instance, one group of religious subjects found repeated satanic messages hidden in recordings when the recordings were played in reverse. What the group was not told is that the recordings were backward renditions of biblical verses.

When subliminal self-help tapes are examined for their effectiveness in controlled studies, they do not produce results beyond chance outcomes. For instance, a famous study in Canada flashed "call now" during TV broadcasts, but no one called. There is no empirical evidence to support that subliminal messages can improve memory, effect purchasing or voting decisions, improve self-esteem, or any of the other claims that sellers of these products declare. Moreover, there is no evidence that subliminal messages can indoctrinate anyone into Satanism or any other lifestyle. Nonetheless, some people still claim that advertisements and music contain hidden messages that force people to behave in certain ways.

Listening to Mozart When You Are Pregnant Can Make Your Kids Smarter

Related to the subliminal learning myth is the myth that babies or children who are exposed to Mozart in the womb or early in their development grow up to be smarter. A company named after Einstein markets many of these products and apparently is a million-dollar franchise. Parents spend tons of dollars on these products because they read that exposure to great art is good for a child's cognitive development. The catch phrase for this is "the Mozart effect."

The myth originated during the 1950s when an ENT doctor claimed that using Mozart's music helped people with speech and hearing disorders. Later in the 1990s, a UC-Irvine study had a group of students listen to a Mozart sonata before taking an IQ test. The effect allegedly raised the IQ scores of the students by an average of eight points. The problem is that an eight-point difference in IQ scores over multiple administrations is not significant and could be an expected practice effect when IQ tests are administered in a relatively short period of time, or it could reflect normal variation in scores of a test administered on different occasions (these are both statistically plausible explanations for the variation of scores seen in this study, and such variation would not be due to the music). The psychologist administering the study clearly was not well trained in test theory.

Nonetheless, the "Mozart effect" was born, and marketers trademarked the phrase "Mozart Effect" and created a whole line of merchandise. Believe it or not, several states even offered funding to develop programs to play classical music to children. Other companies have gone on to claim that listening to Mozart can improve your health.

There has been some evidence to suggest that learning to play an instrument improves self-confidence, concentration, and motor coordination, but even this evidence has been inconsistent when applied to being associated with increased grades in school. Learning to play an instrument *does* provide a person with a form of relaxation and an outlet to deal with stress.

Here is the punch line: One researcher from the original study has stated that he never claimed listening to classical music made the original subjects smarter; instead it was associated with an increase in performance on spatial tasks (which are the tasks in IQ tests that are most subject to short-term practice effects and more variation over repeated testing). Despite numerous efforts to demonstrate a relationship between listening to Mozart, or to classical music in general, and improved cognition, there is currently *no* scientific evidence to indicate that listening to any classical music, country music, rap, or any other type of music improves cognition. Listening to Mozart certainly is not bad for you, but it will not make you or your children smarter.

Some People Have ESP

The term ESP (Extra Sensory Perception) was coined by J. B. Rhine in the 1920s. Rhine investigated paranormal phenomena while at Duke University. ESP refers to a number of psychic abilities, such as telepathy, precognition, clairvoyance (remote viewing), or clairaudience (hearing voices or thoughts psychically). One of the more popular methods to demonstrate that someone has ESP is called the Ganzfeld procedure.

In the Ganzfeld method, one person is a "sender" and another is a "receiver." The sender views randomly chosen pictures or images while the receiver sits in a soundproof chamber with his eyes covered, wearing headphones that play continuous white noise, with a red light shining in the room. The sender concentrates on the image, and the receiver attempts to connect to this image mentally. When ready, the receiver removes the eye covers and picks the image the sender relayed to him from one of four images he is shown. Using this method Bem and Honorton in 1994 reported that their subjects produced a hit rate of 33 percent. Since 25 percent would be the rate expected by chance (one of four), the researchers cited this as evidence for ESP.

But wait a second! This conclusion is based on a lack of knowledge concerning chance. What was observed is known as the *clustering illusion* that commonly occurs in short sequences of random events. For instance, the probability that a single coin toss will

How many times have you thought of a song and then turned on the radio and there it was? How about instances like that? Those demonstrate ESP, right? Well, the problem is that you never consider all the thousands of times you thought of a song and it ***was not*** playing on the radio. What you are falling prey to is known as a ***confirmation bias***.

turn up heads is one in two; however, the probability of getting four straight heads in a row is one in sixteen, which is not all that improbable (this is a simple probability calculation learned by every high school math student). Moreover, even after getting a string of four heads, if one were to keep repeating many sequences of four tosses and averaging the results over thousands of trials, one would produce results consistent with getting heads 50 percent of the time. Yet, many would claim that a single string of four heads in a row is some sort of "hot streak" that falls outside the realm of chance. This is simply not true.

In response to the aforementioned findings, researchers conducted a meta-analysis of thirty controlled studies from seven different facilities that used the Ganzfeld method. Meta-analysis is considered a more powerful statistical procedure than single-study designs, because it combines the results of many studies. The results of the meta-analysis failed to confirm that the hit rate in all of these studies was above chance. Moreover, in reanalyzing Bem and Honorton's results, it was found that the 33 percent hit rate they claimed was not statistically significant.

Susan Blackmore is a researcher in England who was at first a believer in ESP, but after many years of research found no evidence for it. She ran a series of controlled studies on the Ganzfeld method and never found any evidence for ESP. Her conclusions, along with those of many others, indicate that the research supporting ESP is flawed because:

1. The subjects who choose correctly above chance never can tell when they choose correctly or when they do not choose correctly. If ESP is really present in these people, it would be an unconscious process, which is inconsistent with what psychics and supporters claim.

2. Subjects who score above chance are not able to repeat their performance either in different studies or on different occasions. This is more consistent with a cluster illusion or normal runs of hits that occur in chance events.

3. Finally, larger studies and meta-analyses consistently find that there is not better-than-chance accuracy in these studies. Therefore no ESP ability is demonstrated.

GLOSSARY

action potential
The term for the electrical signal in a neuron

afferent nerve cells
Cells that process incoming information

amygdala
A subcortical structure involved in emotion and emotionally charged memories

anterior
Toward the front

apraxia
A dysfunction of movement to command

aphasia
A dysfunction of language

association cortex
Cortical area that receives input from a primary sensory cortex and other areas

autonomic nervous system
The division of the peripheral nervous system controlling involuntary functions

axon
The portion of the neuron that sends a signal

basal ganglia
Several subcortical structures involved in movement

cerebellum
The structure of the posterior portion of the brain involved in movement

central nervous system (cns)
The brain and spinal cord

corpus callosum
The major commissure, or tract, that connects the left and right hemispheres of the brain

cortex
The covering of the brain

cranial nerves
Twelve sets of nerves that do not pass through the spinal cord. These nerves are involved in basic functions.

dementia
A deterioration of intellectual skills

efferent nerve cells
Nerve cells that send information out of the CNS

executive functions
Planning, executing, and inhibiting actions

dendrite
The portion of the neuron that receives information

dorsal
The back side

frontal lobe
The most anterior lobe of the brain, involved in decision making and multiple functions

glial cells
Cells in the CNS that perform a number of maintenance and support functions

gyrus
A wrinkle in the brain tissue

hippocampus
A subcortical brain structure involved in developing new memories

homeostasis
An acceptable range for various bodily states and functions

hormones
Chemicals released from glands into the circulatory system to affect behavior

hypothalamus
The subcortical structure involved in regulating the pituitary gland

inferior
Below

lateral
On the side

lateralization
Describes the situation where the neurons on one side of the brain control the functions on the opposite body side

limbic system
A system of structures in the brain involved in memory and emotion

medial
Toward the middle or inner portion

meninges
Three layers of brain covering: the pia mater, arachnoid, and dura mater

myelin
The fatty sheath covering the axon that facilitates neural transmission

neurons
The nerve cells in the CNS

neurotransmitters
The chemicals that are used in neural communication in the CNS

occipital lobe
The lobe at the posterior portion of the brain, primarily involved in vision

parasympathetic nervous system
A division of the autonomic nervous system that is involved in slowing down involuntary processes

parietal lobe
The lobe of the brain involved in touch and spatial information

perception
The process of interpreting sensory information

peripheral nervous system
Any nervous tissue outside of the CNS

pituitary gland
The master gland in the body

sensation
The detection of physical energy and the environment

sulcus
An indentation in the brain

superior
Above

sympathetic nervous system
A division of the autonomic nervous system that speeds up involuntary processes

synapse
The space between neurons where neurotransmitters are released

thalamus
The subcortical structure that acts as a relay station for all sensory information

transduction
The process of converting physical energy into neural signals

ventral
The belly side

ventricles
The spaces in the brain and spinal cord filled with cerebral spinal fluid

working memory
Analogous to short-term memory. The combination of attention and memory where information is manipulated for very short periods.

REFERENCES

Here are several websites that you may find interesting.

The Brain from Top to Bottom
A good site to see neuroimaging techniques
http://thebrain.mcgill.ca/flash/capsules/outil_bleu13.html#1

Loyola Medical University
Cranial nerves
www.meddean.luc.edu/lumen/MedEd/grossanatomy/h_n/cn/cn1/mainframe.htm

The joy of visual perception
A really good site to investigate afterimages and visual illusions
www.yorku.ca/eye/toc.htm

Apparelyzed
A good site for viewing the dermatomes and spinal cord functioning
www.apparelyzed.com/myo-dermatomes.html

Medscape reference
A depiction of EEG recordings of sleep stages
http://emedicine.medscape.com/article/1140322-overview

Hatton lab projects
Videos of neural migration
www.rockefeller.edu/labheads/hatten/cellmigration.html

Want to Know More?

Resources/References for Chapter 1

Bremner, J. Douglas. *Brain Imaging Handbook* (New York, NY: W. W. Norton, 2005).

Letivan, Irwin B., and Leonard K. Kaczmarek. *The Neuron: Cell and Molecular Biology* (New York, NY: Oxford University Press, 2001).

Shinji Nishimoto, An T. Vu, Thomas Naselaris, Yuval Benjamini, Bin Yu, and Jack L. Gallant, "Reconstructing Visual Experiences from Brain Activity Evoked by Natural Movies," *Current Biology*, 11 October 2011, pp. 1641–1646.

Resources/References for Chapter 2

Kandel, Eric R., James H. Schwartz, Thomas M. Jessell, Steven A. Siegelbaum, and A. J. Hudspeth, eds. *Principles of Neural Science*, 5th ed. (New York, NY: McGraw-Hill, 2012).

Springer, Sally P., and Georg Deutsch. *Left Brain, Right Brain: Perspectives from Cognitive Neuroscience* (New York, NY: W. H. Freeman, 2001).

Resources/References for Chapter 3

Farah, Martha J. *The Cognitive Neuroscience of Vision* (Malden, MA: Blackwell Publishers, 2000).

Robinson, J. O. *The Psychology of Visual Illusion* (Mineola, NY: Dover, 1998).

Resources/References for Chapter 4

Dehaene, Stanislas. *Reading in the Brain: The Science and Evolution of a Human Invention* (New York, NY: Viking, 2009).

Harley, Trevor A. *The Psychology of Language: From Data to Theory*, 3rd ed. (New York, NY: Psychology Press, 2008).

LaPointe, Leonard L. *Aphasia and Related Neurogenic Language Disorders*, 4th ed. (New York, NY: Thieme, 2011).

Moore, Brian C. J. *An Introduction to the Psychology of Hearing*, 6th ed. (Bingley, UK: Emerald Group, 2012).

Snowling, Margaret, and Charles Hulme, eds. *The Science of Reading: A Handbook* (Malden, MA: Blackwell, 2005).

Resources/References for Chapter 5

Melzack, Ronald, and Joel Katz. "The Gate Control Theory: Reaching for the Brain." In *Pain: Psychological Perspectives*, edited by Thomas Hadjistavropoulos and Kenneth D. Craig, pp. 13–34 (Mahwah, NJ: Lawrence Erlbaum Associates, 2004).

Ronald Melzack and Patrick D. Wall, "Pain Mechanisms: A New Theory," *Science*, 19 November 1965, pp. 971–979.

Ramachandran, V. S., and Sandra Blakeslee. *Phantoms in the Brain: Probing the Mysteries of the Human Mind* (New York, NY: William Morrow, 1999).

W. D. Willis, Jr., "The Somatosensory System, with Emphasis on Structures Important for Pain," *Brain Research Reviews*, 12 June 2007, pp. 297–313.

Resources/References for Chapter 6

Beauchamp, Gary K., and Linda Bartoshuk, eds. *Tasting and Smelling: Handbook of Perception and Cognition* (San Diego, CA: Academic Press, 1997).

Rouby, Catherine, Benoist Schaal, Danièle Dubois, Rémi Gervais, and A. Holley, eds. *Olfaction, Taste, and Cognition* (New York, NY: Cambridge University Press, 2002).

Resources/References for Chapter 7

Calais-Germain, Blandine. *Anatomy of Movement*, rev. ed. (Seattle, WA: Eastland Press, 2007).

Edwards, Mark, Kailash Bhatia, Niall Quinn, and Leslie Swinn. *Parkinson's Disease and Other Movement Disorders* (New York, NY: Oxford University Press, 2008).

Enoka, Roger M. *Neuromechanics of Human Movement*, 4th ed. (Champaign, IL: Human Kinetics, 2008).

Resources/References for Chapter 8

Schulkin, Jay. *Rethinking Homeostasis: Allostatic Regulation in Physiology and Pathophysiology* (Cambridge, MA: MIT Press, 2003).

Young, John K. *Hunger, Thirst, Sex, and Sleep: How the Brain Controls Our Passions* (Lanham, MD: Rowman & Littlefield, 2012).

Resources/References for Chapter 9

Dixson, Alan F. *Sexual Selection and the Origins of Human Mating Systems* (New York, NY: Oxford University Press, 2009).

Melissa Hines, "Gender Development and the Human Brain," *Annual Review of Neuroscience,* July 2011, pp. 69–88.

Omoto, Allen, and Howard S. Kurtzman, eds. *Sexual Orientation and Mental Health: Examining Identity and Development in Lesbian, Gay, and Bisexual People (*Washington, DC: American Psychological Association, 2006).

Kolja Schiltz, Joachim Witzel, Georg Northoff, Kathrin Zierhut, Ubo Gubka, Hermann Fillmann, Jörn Kaufmann, Claus Tempelmann, Christine Wiebking, and Bernhard Bogerts, "Brain Pathology in Pedophilic Offenders: Evidence of Volume Reduction in the Right Amygdala and Related Diencephalic Structures," *Archives of General Psychiatry,* June 2007, pp. 737–746.

Shuster, Stephen M., and Michael John Wade. *Mating Systems and Strategies* (Princeton, NJ: Princeton University Press, 2003).

Resources/References for Chapter 10

Berry, Richard B. *Fundamentals of Sleep Medicine: Expert Consult* (Philadelphia, PA: Saunders, 2012).

Committee on Sleep Medicine and Research, Harvey R. Colten, and Bruce M. Altevogt, eds. *Sleep Disorders and Sleep Deprivation: An Unmet Public Health Problem* (Washington, DC: National Academies Press, 2006).

Stickgold, Robert, and Matthew P. Walker. *The Neuroscience of Sleep* (Burlington, MA: Academic Press, 2009).

Resources/References for Chapter 11

Nelson, Charles A., Michelle de Haan, and Kathleen M. Thomas. *Neuroscience of Cognitive Development: The Role of Experience and the Developing Brain (*Hoboken, NJ: Wiley, 2006).

Sprenger, Marilee. *The Developing Brain: Birth to Age Eight* (Thousand Oaks, CA: Corwin Press, 2008).

Stiles, Joan. *The Fundamentals of Brain Development: Integrating Nature and Nurture* (Cambridge, MA: Harvard University Press, 2008).

Resources/References for Chapter 12

Fuster, Joaquín M. *The Prefrontal Cortex*, 4th ed. (London, UK: Academic Press, 2008).

Goldberg, Elkhonon. *The New Executive Brain: Frontal Lobes in a Complex World* (New York, NY: Oxford University Press, 2009).

Levine, Brian, and Fergus I. M. Craik, eds. *Mind and the Frontal Lobes: Cognition, Behavior, and Brain Imaging* (New York, NY: Oxford University Press, 2012).

Resources/References for Chapter 13

Baddeley, Alan. *Working Memory, Thought, and Action* (New York, NY: Oxford University Press, 2007).

Klingberg, Torkel. *The Overflowing Brain: Information Overload and the Limits of Working Memory* (New York, NY: Oxford University Press, 2009).

Posner, Michael I. *Cognitive Neuroscience of Attention*, 2nd ed. (*New York, NY: Guilford Press, 2012).

Resources/References for Chapter 14

Atkinson, R. C., and R. M. Shiffrin. "Human Memory: A Proposed System and Its Control Processes." In *The Psychology of Learning and Motivation,* Vol. 2, edited by Kenneth W. Spence and Janet Taylor Spence, pp. 89–195 (New York, NY: Academic Press, 1967).

Kandel, Eric. *In Search of Memory: The Emergence of a New Science of Mind* (New York, NY: Norton, 2006).

Schacter, Daniel L. *The Seven Sins of Memory: How the Mind Forgets and Remembers* (New York, NY: Houghton Mifflin, 2002).

Tulving, E. "Episodic and Semantic Memory." In *Organization of Memory*, edited by E. Tulving and W. Donaldson, pp. 381–402 (New York, NY: Academic Press, 1972).

Endel Tulving and Stephen A. Madigan, "Memory and Verbal Learning," *Annual Review of Psychology,* February 1970, pp. 437–484.

Resources/References for Chapter 15

Barrett, Lisa Feldman, Paula M. Niedenthal, and Piotr Winkielman, eds. *Emotion and Consciousness (*New York, NY: Guilford Press, 2005).

Friedman, Matthew J., Terence M. Keane, and Patricia A. Resick, eds. *Handbook of PTSD: Science and Practice* (New York, NY: Guilford Press, 2007).

Lazarus, Richard S. *Psychological Stress and the Coping Process* (New York, NY: McGraw-Hill, 1966).

LeDoux, Joseph E. *The Emotional Brain: The Mysterious Underpinnings of Emotional Life* (New York, NY: Simon & Schuster, 1996).

Selye, Hans. *The Stress of Life* (New York, NY: McGraw-Hill, 1976).

Soreq, Hermona, Alon Friedman, and Daniela Kaufer, eds. *Stress: From Molecules to Behavior: A Comprehensive Analysis of the Neurobiology of Stress Responses* (Hoboken, NJ: Wiley-Blackwell, 2009).

Resources/References for Chapter 16

Gazzaniga, Michael S. *Nature's Mind: Biological Roots of Thinking, Emotions, Sexuality, Language, and Intelligence* (New York, NY: Basic Books, 1994).

Geary, David C. *The Origin of Mind: Evolution of Brain, Cognition, and General Intelligence* (Washington, DC: American Psychological Association, 2004).

Sternberg, Robert J., and Scott Barry Kaufman, eds. *The Cambridge Handbook of Intelligence* (New York, NY: Cambridge University Press, 2011).

Resources/References for Chapter 17

Kagan Jerome. *Galen's Prophecy: Temperament in Human Nature* (Boulder, CO: Westview Press, 1998).

Larsen, Randy J., and David M. Buss. *Personality Psychology: Domains of Knowledge about Human Nature* (New York, NY: McGraw-Hill, 2008).

Marcel Zentner and John E. Bates, "Child Temperament: An Integrative Review of Concepts, Research Programs, and Measures," *European Journal of Developmental Science*, 2008, Vol. 2, No. 1/2, pp. 7–37.

Resources/References for Chapter 18

Erickson, Carlton K. *The Science of Addiction: From Neurobiology to Treatment* (New York, NY: W. W. Norton, 2007).

Kuhn, Cynthia M., and George F. Koob, eds. *Advances in the Neuroscience of Addiction* (Boca Raton, FL: CRC Press, 2010).

Maté, Gabor. *In the Realm of Hungry Ghosts: Close Encounters with Addiction* (Berkeley, CA: North Atlantic Books, 2009).

Resources/References for Chapter 19

Bourgeois, Michelle S., and Ellen M. Hickey. *Dementia: From Diagnosis to Management–A Functional Approach* (New York, NY: Psychology Press, 2009).

Ghaemi, S. Nassir. *Mood Disorders: A Practical Guide* (Philadelphia, PA: Lippincott Williams & Wilkins, 2008).

Grandin, Temple. *Thinking in Pictures: My Life with Autism*, expanded ed. (New York, NY: Vintage, 2006).

Torrey, E. Fuller. *Surviving Schizophrenia: A Manual for Families, Consumers, and Providers* (New York, NY: Harper Perennial, 2006).

Resources/References for Chapter 20

Adrian Bangerter and Chip Heath, "The Mozart Effect: Tracking the Evolution of a Scientific Legend," *British Journal of Social Psychology*, 18 December 2002, pp. 605–623.

Susan Blackmore, "What Can the Paranormal Teach Us about Consciousness?" *Skeptical Inquirer*, March/April 2001, pp. 22–27.

Loftus, Elizabeth F., and Katherine Ketchum. *The Myth of Repressed Memory: False Memories and Allegations of Sexual Abuse* (New York, NY: St. Martin's Press, 1996).

Ost, James. "Recovered Memories." In *Handbook of Psychology of Investigative Interviewing: Current Developments and Future Directions*, edited by Ray Bull, Tim Valentine, and Tom Williamson, pp. 181–204 (Chichester, UK: Wiley-Blackwell, 2009).

Vokey, John R. "Subliminal Messages." In *Psychological Sketches*, 6th ed., edited by John R. Vokey and Scott W. Allen, pp. 223–246 (Lethbridge, Alberta: Psyence Ink, 2002).

INDEX